Network Coding at Different Layers
In Wireless Networks

Yang Qin

Editor

Network Coding at Different Layers In Wireless Networks

 Springer

Editor
Yang Qin
Computer Science Department
Shenzhen Graduate School
Harbin Institute of Technology
Xili, Shenzhen, China

ISBN 978-3-319-80650-1   ISBN 978-3-319-29770-5   (eBook)
DOI 10.1007/978-3-319-29770-5

Printed on acid-free paper

This Springer imprint is published by Springer Nature
The registered company is Springer International Publishing AG Switzerland

Preface

Since network coding was proposed by R. Ahlswede, N. Cai, S.Y. Li and R.W. Yang in 2000, it has been widely used in wired networks, wireless networks, p2p content distribution, distributed file storage, network security and other fields. Network coding has been proven to improve network performance by increasing the throughput and decreasing the delay in networks.

Y. Zhu, B. Li and J. Guo deployed network coding in application layer overlay networks. It was one of the earliest applications of network coding in network systems. Network coding was applied in a multicast process to improve end-to-end throughput in the application layer. Recently, Christos Gkantsidis et al. deployed network coding in large-scale content distribution systems; the relevant works were presented in IEEE INFOCOM2015. Various works using network coding have been proposed to enhance the performance of networks. There are several interesting deployments of network coding especially in wireless networks.

This edited book is a collection of valuable contributions from many experienced scientists in the field. It is intended to be a reference book to address the basic concept of network coding, and its typical applications in network systems for both industry and academia. This book aims to introduce the applications of network coding in network systems, especially in wireless network systems, at different layers. It serves as an introductory book for students to gain fundamental knowledge of various aspects of network coding and their applications. It also serves as a rich reference for researchers and engineers to understand the recent technology of network coding.

The book is organized as follows:

Chapter 1 addresses network coding deployments at the physical layer. These deployments significantly improve the quality of signal received. A physical layer wireless network coding scheme is referred to as soft network coding (SoftNC), where the relay nodes apply symbol-by-symbol soft decisions on the received signals from the two end nodes to come up with the network-coded information to be forwarded. According to measures of the soft information adopted, two kinds of SoftNC are proposed in this chapter: amplify-and-forward SoftNC (AF-SoftNC) and soft-bit-forward SoftNC (SBF-SoftNC).

Chapter 2 studies the performance of the data link layer protocol with network coding deployment on the throughput of network coding nodes. The authors discuss two typical data link layer protocols: go-back-N GBN_ARQ and selective repeat SR_ARQ. Their research demonstrates the impact of the number of incoming links to a network coding node on the throughput of network nodes.

Chapter 3 addresses the network coding application at the network layer. Network coding has been implemented with a routing scheme to enhance the throughput. Two methods to adopt network coding are presented here: intra-flow and inter-flow. The intra-flow network coding scheme improves network performance by enhancing the reliability of transmission, while the inter-flow scheme improves performance by enhancing the efficiency of transmission. Readers can also learn how to deploy network coding at the network layer from the recent research works on network coding with multicast.

Chapter 4 presents typical network code research works at the transport layer. Typical works are relevant with well-known protocol TCP. Since TCP is widely used in wired and wireless networks, it could provide end-to-end connection. This chapter introduces a new mechanism for TCP based on network coding, which only requires minor changes to the protocol to achieve incremental deployment. The basic concept of the mechanism is to transmit a linear combination of original packets in the congestion window and simultaneously generate redundant combinations to mask random losses from TCP.

Chapter 5 addresses network coding at application layer multicast. Several multicast schemes with network coding are introduced. This chapter emphasizes that, with peer-to-peer networks at the application layer, network topology can be easily tailored to facilitate network coding. A file sharing system is presented. A network coding scheme for a peer-to-peer multimedia system has been introduced as well. This chapter also discusses the advantages of such schemes.

I would like to thank all the authors for their valuable contributions, profound knowledge and great efforts in the preparation of this book. I would also like to thank the publisher of the book and Ms. Brinda Megasyamalan, Project Coordinator and Ms. Mary E. James, Publishing editor for their patience, support and help.

Xili, China   Yang Qin

Preface

Since network coding was proposed by R. Ahlswede, N. Cai, S.Y. Li and R.W. Yang in 2000, it has been widely used in wired networks, wireless networks, p2p content distribution, distributed file storage, network security and other fields. Network coding has been proven to improve network performance by increasing the throughput and decreasing the delay in networks.

Y. Zhu, B. Li and J. Guo deployed network coding in application layer overlay networks. It was one of the earliest applications of network coding in network systems. Network coding was applied in a multicast process to improve end-to-end throughput in the application layer. Recently, Christos Gkantsidis et al. deployed network coding in large-scale content distribution systems; the relevant works were presented in IEEE INFOCOM2015. Various works using network coding have been proposed to enhance the performance of networks. There are several interesting deployments of network coding especially in wireless networks.

This edited book is a collection of valuable contributions from many experienced scientists in the field. It is intended to be a reference book to address the basic concept of network coding, and its typical applications in network systems for both industry and academia. This book aims to introduce the applications of network coding in network systems, especially in wireless network systems, at different layers. It serves as an introductory book for students to gain fundamental knowledge of various aspects of network coding and their applications. It also serves as a rich reference for researchers and engineers to understand the recent technology of network coding.

The book is organized as follows:

Chapter 1 addresses network coding deployments at the physical layer. These deployments significantly improve the quality of signal received. A physical layer wireless network coding scheme is referred to as soft network coding (SoftNC), where the relay nodes apply symbol-by-symbol soft decisions on the received signals from the two end nodes to come up with the network-coded information to be forwarded. According to measures of the soft information adopted, two kinds of SoftNC are proposed in this chapter: amplify-and-forward SoftNC (AF-SoftNC) and soft-bit-forward SoftNC (SBF-SoftNC).

Chapter 2 studies the performance of the data link layer protocol with network coding deployment on the throughput of network coding nodes. The authors discuss two typical data link layer protocols: go-back-N GBN_ARQ and selective repeat SR_ARQ. Their research demonstrates the impact of the number of incoming links to a network coding node on the throughput of network nodes.

Chapter 3 addresses the network coding application at the network layer. Network coding has been implemented with a routing scheme to enhance the throughput. Two methods to adopt network coding are presented here: intra-flow and inter-flow. The intra-flow network coding scheme improves network performance by enhancing the reliability of transmission, while the inter-flow scheme improves performance by enhancing the efficiency of transmission. Readers can also learn how to deploy network coding at the network layer from the recent research works on network coding with multicast.

Chapter 4 presents typical network code research works at the transport layer. Typical works are relevant with well-known protocol TCP. Since TCP is widely used in wired and wireless networks, it could provide end-to-end connection. This chapter introduces a new mechanism for TCP based on network coding, which only requires minor changes to the protocol to achieve incremental deployment. The basic concept of the mechanism is to transmit a linear combination of original packets in the congestion window and simultaneously generate redundant combinations to mask random losses from TCP.

Chapter 5 addresses network coding at application layer multicast. Several multicast schemes with network coding are introduced. This chapter emphasizes that, with peer-to-peer networks at the application layer, network topology can be easily tailored to facilitate network coding. A file sharing system is presented. A network coding scheme for a peer-to-peer multimedia system has been introduced as well. This chapter also discusses the advantages of such schemes.

I would like to thank all the authors for their valuable contributions, profound knowledge and great efforts in the preparation of this book. I would also like to thank the publisher of the book and Ms. Brinda Megasyamalan, Project Coordinator and Ms. Mary E. James, Publishing editor for their patience, support and help.

Xili, China   Yang Qin

Contents

Chapter 1
Soft Network Coding in Wireless Relay Channels

Zhang Shengli and Zhu Yu

Abstract In traditional designs of applying network coding in wireless two-way relay channels, network coding operates at upper layers above (including) the link layer and it requires the input packets to be correctly decoded at the physical layer. Different from that, this chapter investigates a physical layer wireless network coding scheme, which is referred to as soft network coding (SoftNC), where the relay nodes apply symbol-by-symbol soft decisions on the received signals from the two end nodes to come up with the network-coded information to be forwarded. We do not assume further channel coding on top of SoftNC at the relay node (channel coding is assumed at the end nodes). According to measures of the soft information adopted, two kinds of SoftNC are proposed: amplify-and-forward SoftNC (AF-SoftNC) and soft-bit-forward SoftNC (SBF-SoftNC).

1.1 Introduction

Traditionally, network coding is regarded as a higher-layer technique and applied in operates at upper layers above (including) the link layer. Physical layer network coding PNC [1] is a well-known physical layer network coding scheme with very good performance. Although it is promising from both communication theory and information theory point of view, its implementation is not so straightforward in the current stage of technology development. In this chapter, we discuss another scheme, soft network coding (SoftNC), which can be regarded as an extension of straightforward network coding (SNC) scheme to the real-valued signal and could be easily implemented based on today's technology.

Initially, the research community simply regarded relay protocols in two-way relay channel (TWRC) as a generalization of the protocols of one-way channel

Z. Shengli (✉)
School of Information Engineering, Shenzhen University, Shenzhen, China
e-mail: zsl@szu.edu.cn

Z. Yu
Department of Communication Science and Engineering, Fudan University, Shanghai, China
e-mail: zhuyu@fudan.edu.cn

© Springer International Publishing Switzerland 2016
Y. Qin (ed.), *Network Coding at Different Layers In Wireless Networks*,
DOI 10.1007/978-3-319-29770-5_1

1

(OWRC) [2, 3]. With the application of network coding in TWRC [4], in which the SNC scheme was proposed, new possibilities have been opened up. However, previous designs of SNC require correct channel decoding of the received packets from the two ends at the relay node, which may limit the throughput of TWRC. This is similar to that in OWRC, where the performance of the decode-and-forward protocol may be much worse than the performance of the amplify-and-forward protocol under certain scenarios [5]. Furthermore, due to the time variations of the channel fading, it cannot be always assumed that the received packet is decoded correctly, especially when the channel is in deep fading. In addition, in some situations, power consumption at the relay node is a concern (e.g., the relay node is a normal user with limited battery power) and the channel decoding processing may consume excessive amount of power.

In this chapter, to remove the requirement of channel decoding, we propose a new wireless network coding scheme, referred to as soft network coding (SoftNC), where the relay node applies symbol-by-symbol soft decisions on the received signals from the two end nodes to come up with the network-coded information to be forwarded. Note that channel coding is only performed at the end nodes but not the relay node. In particular, the relay node does not perform channel decoding and re-encoding and channel coding is on an end-to-end basis where only the end nodes are involved in channel coding and decoding. In SoftNC, the forwarded signal is actually the soft information of the bits that are obtained by doing the XOR operation to the two code words received, respectively, from the two end nodes. According to measures of soft information adopted, two kinds of SoftNC are proposed: amplify-and-forward SoftNC (AF-SoftNC) and soft-bit-forward SoftNC (SBF-SoftNC). In the former, the log-likelihood ratios (LLR) of the bits are generated and forwarded; in the latter, the soft bits (i.e., the MMSE estimation of the XOR-ed bit) are generated and forwarded.

This chapter also analyzes the performance of the two proposed SoftNC schemes in terms of the maximum achievable information rate (MAIR), defined as the ergodic mutual information between the two end nodes. We provide closed-form approximations of the MAIR of the two SoftNC schemes. It is shown that the analytical results are very close to the true simulated information rate that is obtained according to the definition of mutual information. Our simulation shows that AF-SoftNC and SBF-SoftNC can obtain substantial MAIR improvements over the conventional two-way relay protocols with or without network coding. Since the proposed SoftNC design also does not require any channel decoding and re-encoding processing at the relay node, it is a very promising network coding method in terms of actual practice in wireless networks.

Related Work: The fundamental idea behind the proposed SoftNC design is that due to the unreliability of the wireless fading channels instead of forwarding the decoded-and-network-coded (XOR-ed) bit, the relay node can calculate and forward the likelihood information, i.e., how likely the network-coded bit is "0" or "1." The AF-SoftNC design has been considered in a preliminary version of this chapter [6]. The same similar idea was independently proposed in a two sources relay system in [7]. More recently, an encoding-decoding framework and BER analysis in fading channel for the two-source relay system have been considered in [8]. Different from these works, where the soft information is obtained based on the whole received packet (e.g., after the soft-input soft-output channel decoding), our work focuses on the network coding where the relay directly obtains the symbol-by-symbol soft information of the network-coded bit based on the received signals from the two end nodes without any channel coding operation. This greatly reduces the computational complexity at the relay node since the channel decoding processing occupies most of the baseband power.

The rest of this chapter is organized as follows. Section 1.2 presents the system model. In Sect. 1.3, we present two soft network coding designs, AF-SoftNC and SBF-SoftNC. We analyze their MAIR in Sect. 1.4. Section 1.5 presents our numerical simulation results. Section 1.6 concludes this chapter and Appendix 1 provides appending proofs.

1.2   System Model

Consider a two-way relay communication system as shown in Fig. 1.1, where the two end nodes, N_1 and N_2, exchange their information with the help of the relay node N_3. We assume that all the three nodes work in the half-duplex mode, where each node either transmits or receives at a particular time. We also assume that different transmissions among the three nodes are separated in non-overlapping time slots.[1] Due to the broadcast nature of the wireless medium, packets transmitted by any node can be received by the other two nodes. In the first slot, node N_1 sends its packet to node N_2 (the relay node N_3). In the second time slot, node N_2 sends its packet to node N_1 (the relay node N_3).[2] If network coding is used at the relay node N_3, in the third time slot, node N_3 will combine the two packets received in the previous two time slots with network coding and forward the network-coded packet to the other two nodes. If network coding is not used, node N_3 will forward the two received packets in the third and fourth time slots, respectively.

Let $U_i = [u_i[0], \cdots, u_i[n], \cdots, u_i[K_i-1]]$ [3] denote the information packet transmitted by the two end nodes N_i, where $i=1,2$, $u_i[n] \in \{0,1\}$, and K_i is the corresponding packet length. Channel coding (including interleaving) is usually performed for certain transmission reliability in wireless channels. Let Γ_i denote the channel coding scheme at node N_i, and let $D_i = [d_i[0], \cdots, d_i[n], \cdots, d_i[M_i-1]]$

---

[1]This is to guarantee that different transmissions among the nodes are through orthogonal channels. Besides through non-overlapping time slots, they can also be seen as through orthogonal frequency bands or through orthogonal spread spectrum codes. For simplicity, we assume all the time slots have identical time duration.

[2]This chapter first discusses the case without direct link and then extends to the case with direct link.

[3]Throughout this chapter, we use uppercase letters to denote packets and the corresponding lowercase letters to denote the symbols in the packets.

Fig. 1.1 Wireless two-way
relay channels

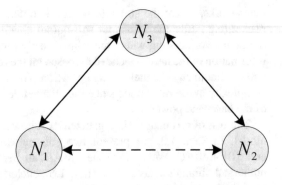

denote the code word, where $d_i[n] \in \{0, 1\}$ and M_i is the codeword length. For simplicity, we assume that the two end nodes use the same coding schemes, i.e., $\Gamma \equiv \Gamma_1 = \Gamma_2$, and the same packet length i.e., $K \equiv K_1 = K_2$, and $M \equiv M_1 = M_2$.[4] We further assume that BPSK modulation is used, and then the relationship between a BPSK symbol and the corresponding coded bit is given by

$$x_i[n] = 1 - 2d_i[n]. \tag{1.1}$$

In the following, we define that Γ_i includes both channel coding and BPSK modulation. The relationship between the information packet and the transmitted BPSK packet can be represented by

$$X_i = \Gamma_i(U_i) \qquad U_i = \Gamma_i^{-1}(X_i) \tag{1.2}$$

where Γ_i^{-1} denotes the decoding processing. Suppose code word length is long and it spans several coherence periods. We could consider the whole code word as being divided into L blocks with the block length Q less than or equal to the length of the channel coherence time (i.e., at least Q symbols are covered during the coherence time). The received signal in the lth block at node N_j can be expressed as

$$y_{i,j}^l[m] = h_{i,j}^l x_i^l[m] + w_{i,j}^l[m] \qquad \text{for } i,j = 1, 2, 3 \text{ and } m = 0, \ldots, Q-1 \tag{1.3}$$

where $x_i^l[m]$ is the mth symbol in the lth block transmitted by node N_i; $y_{i,j}^l[m]$ is the corresponding signal received at node N_j; $w_{i,j}^l[m]$ is the corresponding complex Gaussian noise at node N_j with normalized variance per dimension, i.e., $w_{i,j}^l \sim \mathcal{CN}(0, 2)$; and $h_{i,j}^l$ is the corresponding channel fading coefficient. It should be noted here that throughout this chapter, the transmit power is normalized to one, and $h_{i,j}^l$ actually includes the real transmit power, the path loss effect, and the

---

[4]We will discuss the system design with different channel coding schemes at the two end nodes in part III. 3.

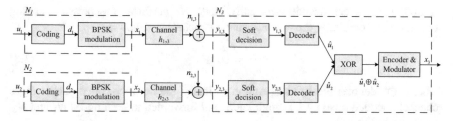

Fig. 1.2 System diagram of the traditional network coding scheme

small-scale multipath fading effect. When the block number is large and a random interleave is applied among the blocks, we can assume that $h_{i,j}^l$ is an independent identical complex Gaussian distributed for different blocks with the variance $\lambda_{ij} = E\{|h_{ij}|^2\}$.

Next, we briefly review the traditional straightforward network coding (SNC) scheme [4], where the network coding operation is performed at the relay node N_3, as shown in Fig. 1.2. Assume that node N_3 has perfect channel state information (CSI) of $h_{i,3}^l$. By performing coherent demodulation to the received signal from the end node N_i, we have

$$\tilde{y}_{i,3}^l[m] \triangleq \mathrm{Re}\left\{\frac{\left(h_{i,3}^l\right)^*}{\left|h_{i,3}^l\right|} y_{i,3}^l[m]\right\} = \left|h_{i,3}^l\right| x_i^l[m] + \tilde{w}_{i,3}^l[m] \tag{1.4}$$

where the superscript * denotes the conjugation and the noise is with the distribution $\tilde{w}_{i,3}^l[m] \sim \mathcal{N}(0, 1)$. Since soft decoding is considered in the system, two kinds of soft information can be generated as the measure of the detection result. These are the log-likelihood ratio (LLR) value

$$v_{i,3}^l[m] = \ln\left(\frac{P\left(x_i^l[m] = 1 \middle| \tilde{y}_{i,3}^l[m]\right)}{P\left(x_i^l[m] = -1 \middle| \tilde{y}_{i,3}^l[m]\right)}\right) = \ln\left(\frac{P\left(\tilde{y}_{i,3}^l[m] \middle| x_i^l[m] = 1\right)}{P\left(\tilde{y}_{i,3}^l[m] \middle| x_i^l[m] = -1\right)}\right)$$

$$= 2\left|h_{i,3}^l\right| \tilde{y}_{i,3}^l[m] \tag{1.5}$$

and the soft bit value [9]

$$v_{i,3}^l[m] = P\left(x_i^l[m] = 1 \middle| \tilde{y}_{i,3}^l[m]\right) - P\left(x_i^l[m] = -1 \middle| \tilde{y}_{i,3}^l[m]\right)$$

$$= \frac{\exp\left(2\left|h_{i,3}^l\right| \tilde{y}_{i,3}^l[m]\right) - 1}{\exp\left(2\left|h_{i,3}^l\right| \tilde{y}_{i,3}^l[m]\right) + 1} = \tanh\left(\left|h_{i,3}^l\right| \tilde{y}_{i,3}^l[m]\right) . \tag{1.6}$$

By sending the soft information to the channel decoder, the decoded packets are given by

$$\hat{U}_1 = \Gamma^{-1}(V_{1,3}) \qquad \hat{U}_2 = \Gamma^{-1}(V_{2,3}). \tag{1.7}$$

If both packets are decoded correctly, node N_3 performs the network coding operation by combining the two information packets as follows:

$$U_3 = \widehat{U}_1 \oplus \widehat{U}_2 \tag{1.8}$$

where "$\oplus$" denotes the XOR operation.[5] Finally, the network-coded packet U_3 is channel encoded (also by Γ), modulated, and forwarded to both nodes N_1 and N_2, as shown in Fig. 1.2.

During the three time slots in one transmission cycle, every end node receives two packets. One packet is received from its counterpart node in either the first or second time slot, and the other packet is from the relay node in the third time slot. The channel decoding processing of the two packets is the same for the two end nodes. Take node N_2 as an example. It receives $Y_{1,2}$ from node N_1 in the first time slot and $Y_{3,2}$ from node N_3 in the third time slot. After removing the self-information in $Y_{3,2}$, N_2 obtains a noise-corrupted code word of U_1, which is referred to as $\hat{Y}_{3,2}$. It can be seen that actually $Y_{1,2}$ and $\hat{Y}_{3,2}$ are the two received independent copies of the code word X_1. N_2 can perform the maximum ratio combination (MRC) to $Y_{1,2}$ and $\hat{Y}_{3,2}$ before channel decoding.

1.3 Soft Network Coding Design

In this section, we first introduce the basic idea of SoftNC and then propose the SoftNC design for practical systems when fading and noise effects are considered. Finally, we discuss the SoftNC design in TWRC when the two end nodes use different channel coding schemes.

1.3.1 Basic Idea

As shown in Sect. 1.1, in the SNC scheme, the network combination is performed after the successful decoding of the packets from the two end nodes. However, due to the wireless channel fading effect, the received packet may not always be decoded successfully. Furthermore, in some situation, for example, when the relay node is a normal mobile user, the battery power is limited and should be used as efficient as possible. However, the channel decoding processing is power hungry, especially when advanced channel codes, such as turbo codes and LDPC codes, are used.

---

[5]It is shown in [18] that the general network coding combination is the linear operation over a finite field. The addition over GF(1.2), i.e., XOR operation, is usually considered in practical networks for its simplicity and good performance [4].

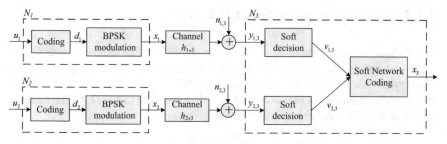

Fig. 1.3 System diagram of the proposed soft network coding scheme

In contrast to the traditional network coding scheme, where network coding is performed after the channel decoding processing, in the proposed scheme, as shown Fig. 1.3, network coding is performed prior to the channel decoding processing by directly combining the soft decisions.

The basic idea stems from the linear property of the channel code. That is, the linear combination of the two code words, which are generated from exactly the same coding scheme with the same length, is actually another code word. This linearity can be formulated as

$$\Gamma\left(U_1 \oplus U_2\right) = \Gamma\left(U_1\right) \oplus \Gamma\left(U_2\right). \tag{1.9}$$

Almost all practical wireless channel codes, such as convolutional codes, turbo codes, and LDPC codes, are linear codes. By applying this property of the channel codes and the fact that network coding is also a linear mapping, it is easily seen that the network coding combination can be done on the code words. This motivates the proposed SoftNC scheme, as shown in Fig. 1.3. By carefully combining the soft decisions $V_{1,3}$ and $V_{2,3}$, the output packet of SoftNC, denoted by V_3, is in fact the code word of the target information packet $U_1 \oplus U_2$. For the simplicity of explanation, if we ignore the noise and fading effects, the SoftNC design can be expressed as

$$\begin{aligned} U_3 &= \Gamma^{-1}\left(V_3\right) = \Gamma^{-1}\left(V_1 \oplus V_2\right) = \Gamma^{-1}\left(D_1 \oplus D_2\right) \\ &= \Gamma^{-1}\left(\Gamma\left(U_1\right) \oplus \Gamma\left(U_2\right)\right) = \Gamma^{-1}\Gamma\left(U_1 \oplus U_2\right) = U_1 \oplus U_2 \end{aligned} \tag{1.10}$$

By comparing SNC in Fig. 1.2 with SoftNC in Fig. 1.3, we see that the relay in SoftNC performs network coding without any channel decoding/encoding process, while the SNC scheme requires two channel decoding processes and one channel encoding process. Since most of the power in baseband signal processing is consumed by the channel decoding, SoftNC can greatly increase the power efficiency of relay nodes in wireless networks.

1.3.2 SoftNC Design in Fading Channels

In practical wireless channels, the data transmission is unavoidably affected by noise and fading. Before considering these effects in SoftNC, we first review the memoryless relay protocols in the well-known OWRC [10]. That is, we just consider the relay direction $N_1 \rightarrow N_3 \rightarrow N_2$ in Fig. 1.1. According to Eq. (1.4), the received packet at N_3 after coherent demodulation is given by

$$\tilde{y}_{1,3}^l [m] = \left| h_{1,3}^l \right| x_1^l [m] + \tilde{w}_{1,3}^l [m]. \tag{1.11}$$

In the literature of OWRC, there are mainly two kinds of memoryless relay protocols [10], where the signal to be forwarded can be expressed as an estimate of $x_1^l[m]$ from the received signal $\tilde{y}_{1,3}^l[m]$. The estimated signals in the two relay protocols correspond to the two soft information measures in Eqs. (1.5) and (1.6), respectively. Specifically:

1. Amplify and forward (AF)

$$v_{1,3}^l [m] = \ln \left(\frac{P\left(x_1^l [m] = 1 \,\middle|\, \tilde{y}_{1,3}^l [m] \right)}{P\left(x_1^l [m] = -1 \,\middle|\, \tilde{y}_{1,3}^l [m] \right)} \right) = 2 \left| h_{1,3}^l \right| \tilde{y}_{1,3}^l [m]. \tag{1.12}$$

   It can be seen from Eq. (1.12) that in the AF protocol the estimated signal is actually the LLR of the transmitted signal from N_1.
2. Soft bit forward (SBF)

$$v_{1,3}^l [m] = P\left(x_1^l [m] = 1 \,\middle|\, \tilde{y}_{1,3}^l [m] \right) - P\left(x_1^l [m] = -1 \,\middle|\, \tilde{y}_{1,3}^l [m] \right) = \tanh\left(\left| h_{1,3}^l \right| \tilde{y}_{1,3}^l [m] \right). \tag{1.13}$$

The SBF protocol was referred to as estimate-and-forward protocol in [10]. It was shown there that SBF is actually a method to perform the MMSE estimation to the source packet at the relay node. It was proved that SBF is optimal in terms of maximizing GSNR (general SNR [10]) of the received signal at the sink and maximizing the mutual information between the source and the sink in the conventional OWRC [11].

Finally, in each protocol, the estimated signal is normalized

$$x_3^l [m] = \sqrt{ \frac{1}{E\left\{ \left(v_{1,3}^l [m] \right)^2 \right\}} }\, v_{1,3}^l [m] \tag{1.14}$$

and forwarded to the destination node N_2.

In SoftNC, the two estimation methods above are extended to TWRC. The difference now is that the relay node estimates $\left(x_1^l [m] \cdot x_2^l [m] \right)$ (rather than individual $x_i^l[m]$) with the knowledge of $\tilde{y}_{1,3}^l[m]$ and $\tilde{y}_{2,3}^l[m]$.

1. AF-SoftNC

As shown in Eq. (1.12), AF in OWRC actually can be seen as forwarding the LLR of the received bits to the destination node. In TWRC, the forwarded signal, i.e., the LLR of the network-coded bit $\left(x_1^l[m] \cdot x_2^l[m]\right)$, is given by

$$
\begin{aligned}
& v_3^l[m] \\
& = \ln \frac{P\left(x_1^l[m] \cdot x_2^l[m]=1 \mid \tilde{y}_{1,3}^l[m], \tilde{y}_{2,3}^l[m]\right)}{P\left(x_1^l[m] \cdot x_2^l[m]=-1 \mid \tilde{y}_{1,3}^l[m], \tilde{y}_{2,3}^l[m]\right)} \\
& = \ln \frac{P\left(x_1^l[m]=1, x_2^l[m]=1 \mid \tilde{y}_{1,3}^l[m], \tilde{y}_{2,3}^l[m]\right)+P\left(x_1^l[m]=-1, x_2^l[m]=-1 \mid \tilde{y}_{1,3}^l[m], \tilde{y}_{2,3}^l[m]\right)}{P\left(x_1^l[m]=1, x_2^l[m]=-1 \mid \tilde{y}_{1,3}^l[m], \tilde{y}_{2,3}^l[m]\right)+P\left(x_1^l[m]=-1, x_2^l[m]=1 \mid \tilde{y}_{1,3}^l[m], \tilde{y}_{2,3}^l[m]\right)} \quad (1.15) \\
& = \ln \left(\frac{1+\exp\left(2\left|h_{1,3}^l\right|\tilde{y}_{1,3}^l[m]+2\left|h_{2,3}^l\right|\tilde{y}_{2,3}^l[m]\right)}{\exp\left(2\left|h_{1,3}^l\right|\tilde{y}_{1,3}^l[m]\right)+\exp\left(2\left|h_{2,3}^l\right|\tilde{y}_{2,3}^l[m]\right)}\right)
\end{aligned}
$$

As widely used in channel decoding algorithms [12], Eq. (1.15) can be further approximated as

$$
v_3^l[m] \approx \text{sign}\left(\tilde{y}_{1,3}^l[m] \cdot \tilde{y}_{2,3}^l[m]\right) \min\left\{2\left|h_{1,3}^l\right|\tilde{y}_{1,3}^l[m], 2\left|h_{2,3}^l\right|\tilde{y}_{2,3}^l[m]\right\}. \quad (1.16)
$$

The mathematics behind this approximation is complicated. An insight of this approximation is that since $v_3^l[m]$ is an estimate of $x_1^l[m]x_2^l[m]$ from $\tilde{y}_{1,3}^l[m]$ and $\tilde{y}_{2,3}^l[m]$, its signature must be $\text{sign}\left(\tilde{y}_{1,3}^l[m] \cdot \tilde{y}_{2,3}^l[m]\right)$ and its belief must not be larger than either $\tilde{y}_{1,3}^l[m]$ or $\tilde{y}_{2,3}^l[m]$.

2. Soft bit forward (SBF)

Based on the definition of soft bit in Eqs. (1.6) and (1.14), in the proposed SBF-SoftNC design in TWRC, the soft bit of $x_1^l[m] \cdot x_2^l[m]$ is defined by

$$
\begin{aligned}
& v_3^l[m] \\
& = P\left(x_1^l[m] \cdot x_2^l[m]=1 \mid \tilde{y}_{1,3}^l[m], \tilde{y}_{2,3}^l[m]\right) - P\left(x_1^l[m] \cdot x_2^l[m]=-1 \mid \tilde{y}_{1,3}^l[m], \tilde{y}_{2,3}^l[m]\right). \\
& \hspace{11cm} (1.17)
\end{aligned}
$$

With further derivation, we have

$$
\begin{aligned}
v_3^l[m] & = \frac{\dfrac{P\left(x_1^l[m] \cdot x_1^l[m]=1 \mid \tilde{y}_{1,3}^l[m], \tilde{y}_{2,3}^l[m]\right)}{P\left(x_1^l[m] \cdot x_1^l[m]=-1 \mid \tilde{y}_{1,3}^l[m], \tilde{y}_{2,3}^l[m]\right)}-1}{\dfrac{P\left(x_1^l[m] \cdot x_1^l[m]=1 \mid \tilde{y}_{1,3}^l[m], \tilde{y}_{2,3}^l[m]\right)}{P\left(x_1^l[m] \cdot x_1^l[m]=-1 \mid \tilde{y}_{1,3}^l[m], \tilde{y}_{2,3}^l[m]\right)}+1} \\
& = \frac{\exp\left(2\left|h_{1,3}^l\right|\tilde{y}_{1,3}^l[m]\right)-1}{\exp\left(2\left|h_{1,3}^l\right|\tilde{y}_{1,3}^l[m]\right)+1} \cdot \frac{\exp\left(2\left|h_{2,3}^l\right|\tilde{y}_{2,3}^l[m]\right)-1}{\exp\left(2\left|h_{2,3}^l\right|\tilde{y}_{2,3}^l[m]\right)+1} \\
& = \tanh\left(\left|h_{1,3}^l\right|\tilde{y}_{1,3}^l[m]\right) \tanh\left(\left|h_{2,3}^l\right|\tilde{y}_{2,3}^l[m]\right).
\end{aligned} \quad (1.18)
$$

It can be seen from Eq. (1.18) that the soft bit of $x_1^l[m] \cdot x_2^l[m]$ is in fact the product of the soft bit of $x_1^l[m]$ and that of $x_2^l[m]$.

Similar to the one-way relay case, the estimate of the network-coded bit $\left(x_1^l[m] \cdot x_2^l[m]\right)$ in Eqs. (1.15) and (1.18) is normalized according to the power constraint at the relay node

$$x_3^l[m] = \alpha v_3^l[m] = \sqrt{\frac{1}{E\left\{\left(v_3^l[m]\right)^2\right\}}} v_3^l[m].$$ (1.19)

Finally, the output of SoftNC, $x_3^l[m]$, is broadcast to both end nodes.

1.4 Performance Analysis of SoftNC without Direct Link

In this section, we analyze the performance of TWRC with SoftNC by investigating the maximum achievable information rate (MAIR), which is defined as the ergodic mutual information that can be reliably transmitted between the two end nodes in the whole relay process.

We first discuss the case that the direct link does not exist. We analyze the channel capacity of the proposed scheme and compare it to that of the traditional approach. The channel discussed here is defined as the equivalent channel whose input is $X_1 * X_2$, with Gaussian additive noise at the receiver. It is assumed that the two transmission channels between the source nodes and the intermediate node are independent and with the same random properties. In this case, the capacity of the equivalent channel is equal to that of the transmission channel for the traditional design.

We now assume that the packets are transmitted in AWGN channels and that SBF scheme is used at the relay node. As we mentioned above, the capacity of the equivalent channel in the traditional design is equal to that of the real transmission channel, which is given as follows according the definition of mutual information

$$R_\mathrm{T} = \sum_{x_{i,n}=\pm 1} \int_{y_{i,n}} p\left(y_{i,n}, x_{i,n}\right) \log_2 \left(\frac{p\left(y_{i,n} \big| x_{i,n}\right)}{\sum_{\tilde{x}=\pm 1} p\left(y_{i,n} \big| \tilde{x}\right)}\right) dy_{i,n}.$$ (1.20)

where $p(y_{i,n}, x_{i,n})$ and $p\left(y_{i,n} \big| x_{i,n}\right)$ denote the joint and the conditional probability density function (PDF) between $y_{i,n}$ and $x_{i,n}$, respectively. By following the same derivation procedure of [13], we have Eq. (1.20) as

$$R_\mathrm{E} = \left\{ \log_2 \left(\frac{2}{1+\exp(-2y_{i,m}/\sigma^2)}\right) \Big| x_{i,m}=1 \right\} = E\left\{ \log_2 \left(\frac{2}{1+\exp(-v_{i,m})}\right) \Big| x_{i,m}=1 \right\}$$ (1.21)

where the expectation operation is taken with respect to $y_{i,n}$.

The capacity of the equivalent channel in the proposed scheme can be obtained by calculating the mutual information between the input signal $X_1 * X_2$ and the output signal V_3. That is

$$R_{NC} = \sum_{\substack{x_1 \\ \overline{=}, n \ *x_{2,n} \\ \pm}} \int_{v_{3,n}} p(v_{3,n}, x_{1,n}*x2, n) \log_2 \left(\frac{p(v_{3,n}|x_{1,n}*x_{2,n})}{\sum_{\tilde{x}_{1,n}*\tilde{x}_{2,n}=\pm 1} p(v_{3,n}|\tilde{x}_{1,n}*\tilde{x}_{2,n})} \right) dv_{3,n}$$

(1.22)

where $v_{3,n}$ is given in Eq. (1.18). It turns out that it is very difficult to obtain the PDF of $v_{3,n}$. However, through simulations, we find that $v_{3,n}$ can be approximated as Gaussian distribution with a larger variance. With this approximation, we have

$$R_{NC} = E \left\{ \log \left(\frac{2}{1 + \exp(-v_{3,n})} \right) \middle| x_{1,n} * x_{2,n} = 1 \right\}$$

(1.23)

The numerical results of the comparison of the capacity between the traditional and the proposed designs are shown as follows.

It is also shown in Fig. 1.4 that the proposed design has lower capacity than the traditional one. This is because the combination prior to the channel coding enhances the noise effect and increases the error probability. However, as seen from Fig. 1.4, as SNR increases, the performance loss will decrease. In the moderate and high region of SNR values, the proposed design can achieve almost the same capacity as the traditional one. Note that the complexity of our SoftNC is much less than the traditional scheme.

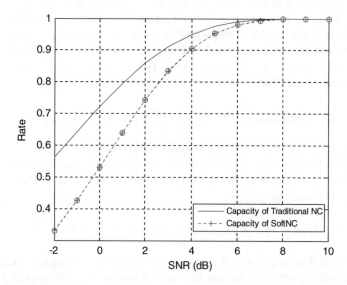

Fig. 1.4 Capacity comparison of the proposed joint network coding-channel decoding design and the traditional network coding design

1.5   Performance Analysis of SoftNC with Direct Link

This section discusses the performance of SoftNC with direct link. First, the ergodic mutual information between the two ends is

$$\mathcal{R} = \underset{h_{i,j}}{E} \left\{ \mathcal{I}_{\left(h_{1,2}, h_{1,3}, h_{3,2}\right)} (x_1; y_{3,2}, y_{1,2}) + \mathcal{I}_{\left(h_{2,1}, h_{2,3}, h_{3,1}\right)} (x_2; y_{3,1}, y_{2,1}) \right\}. \qquad (1.24)$$

where $\mathcal{I}_{\left(h_{1,2}, h_{1,3}, h_{3,2}\right)} (x_1; y_{3,2}, y_{1,2})$ denotes how much information node N_2 can obtain from the received signals $y_{3,2}$ and $y_{1,2}$ given $h_{1,2}$, $h_{1,3}$, and $h_{3,2}$. Likewise, $\mathcal{I}_{\left(h_{2,1}, h_{2,3}, h_{3,1}\right)} (x_2; y_{3,1}, y_{2,1})$ denotes how much information node N_1 can obtain from the received signals $y_{3,1}$ and $y_{2,1}$ given $h_{2,1}$, $h_{2,3}$, and $h_{3,1}$. Note that the subscript in $\mathcal{I}_{(a)}(\cdot)$ denotes the fact that $\mathcal{I}_{(a)}(\cdot)$ is a function of a. As shown in [14], the MAIR $\mathcal{R}$ in Eq. (1.24) is approachable when the capacity-achieving channel coding spans a number of coherent periods to average out both the Gaussian noise and the fluctuations of the channel. That is,

$$\mathcal{R} = \lim_{L \to \infty} \sum_{l=1}^{L} \left(\mathcal{I}_{\left(h_{1,2}^l, h_{1,3}^l, h_{3,2}^l\right)} (x_1; y_{3,2}, y_{1,2}) + \mathcal{I}_{\left(h_{2,1}^l, h_{2,3}^l, h_{3,1}^l\right)} (x_2; y_{3,1}, y_{2,1}) \right). \qquad (1.25)$$

The MAIR analysis consists of three steps. In the first two steps, we calculate the information rate that can be reliably transmitted between the two end nodes with the help of SoftNC at the relay node N_3. Since it has been shown in Sect. 1.2 that it is the information of the network-coded bits $d_1 \oplus d_2$ (or equivalently $x_1 \cdot x_2$) that node N_3 forwards and that the forwarded signal x_3 in SoftNC can be seen as an estimate of $x_1 \cdot x_2$ based on the received signals $y_{1,3}$ and $y_{2,3}$, the information of $d_1 \oplus d_2$ contained in the forwarded signal x_3 can be calculated by assuming a virtual channel whose input is $x_1 \cdot x_2$ and the output is x_3. As shown in Fig. 1.5, in the first step, we calculate the mutual information between the virtual input $x_1 \cdot x_2$ and the forwarded signal x_3. That is,

$$\mathcal{I}_{\text{step1}} \triangleq \mathcal{I}_{\left(h_{1,3}, h_{2,3}\right)} (x_3; (d_1 \oplus d_2)) = \mathcal{I}_{\left(h_{1,3}, h_{2,3}\right)} (x_3; (x_1 \cdot x_2)). \qquad (1.26)$$

Note that $\mathcal{I}_{\text{step1}}$ is a function of the channel coefficients $h_{1,3}$ and $h_{2,3}$.

Based on the result in Eq. (1.26), in the second step, we then calculate the total information which can be exchanged between the two end nodes via the relay node N_3, which is defined as

$$\mathcal{I}_{\text{step2}} \triangleq \mathcal{I}_{\left(h_{1,3}, h_{2,3}, h_{3,2}\right)} (x_1; y_{3,2}) + \mathcal{I}_{\left(h_{1,3}, h_{2,3}, h_{3,1}\right)} (x_2; y_{3,1}). \qquad (1.27)$$

Finally, in the third step, the total mutual information can be obtained by combining the result in the second step and the information that can be exchanged in the direct link between the two end nodes. In the following we analyze the MAIR performance of AF-SoftNC and SBF-SoftNC protocols, respectively.

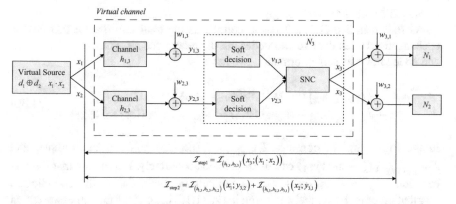

Fig. 1.5 Mutual information notations in the first and second steps in the analysis for the proposed AF-SoftNC and SBF-SoftNC schemes

1.5.1 Analysis of AF-SoftNC

Step 1: *Bounds of $\mathcal{I}_{\text{step1}}$*

As shown in Eq. (1.15), the output of AF-SoftNC, x_3, is the LLR of $x_1 \cdot x_2$ based on the received signals $y_{1,3}$ and $y_{2,3}$. It turns out that the distribution of x_3 is very complicated and it is hard to calculate $\mathcal{I}_{\text{step1}}$ directly. However, an upper bound and a lower bound of $\mathcal{I}_{\text{step1}}$ can be found.

Theorem 1. For two Gaussian channels with any given coefficients $h_{1,3}$ and $h_{2,3}$, the mutual information $\mathcal{I}_{\text{step1}}$ in the AF-SoftNC and SBF-SoftNC protocols is bounded by

$$C_{\text{BPSK}}(h_{1,3}) \cdot C_{\text{BPSK}}(h_{2,3}) \leq \mathcal{I}_{\text{step1}} \leq \mathcal{G}\left(C_{\text{BPSK}}(h_{1,3}), C_{\text{BPSK}}(h_{2,3})\right) \quad (1.28)$$

where $C_{\text{BPSK}}(h) \triangleq \mathcal{I}(x; (y = hx + w))$ is the mutual information of a Gaussian channel with BPSK modulated input x, channel coefficient h, and Gaussian noise $w \sim \mathcal{N}(0, 1)$. Its numerical calculation can be referred to the *Appendix* of [15]. The function $\mathcal{G}(x, y)$ in Eq. (1.28) is defined as

$$\mathcal{G}(x, y) \triangleq 1 - H\left(H^{-1}(x)\left(1 - H^{-1}(y)\right) + H^{-1}(y)\left(1 - H^{-1}(x)\right)\right) \quad (1.29)$$

where $H(p) \triangleq -p \log p - (1 - p) \log (1 - p)$ is the entropy of binary distribution with probability p and $H^{-1}(\cdot)$ is the inverse function of $H(\cdot)$.

The proof of this theorem is given in Appendix 1, where it is shown that Theorem 1 can be seen as a corollary of Theorem 2 in [16]. According to the results in [16] and our extensive simulations, the two bounds are close to each other and both of them can serve as good approximations to $\mathcal{I}_{\text{step1}}$. Some simulation result about Eq. (1.28) will be given later.

Step 2: *Approximation of* $\mathcal{I}_{\text{step2}}$

As shown in Eq. (1.27), $\mathcal{I}_{\text{step2}}$ is defined as the total information that can be exchanged between the two end nodes via the relay node N_3. Since in TWRC each end node knows its self-information, we have

$$
\begin{aligned}
\mathcal{I}_{\text{step2}} &\triangleq \mathcal{I}_{(h_{1,3},h_{2,3},h_{3,2})}\,(x_1; y_{3,2}) + \mathcal{I}_{(h_{1,3},h_{2,3},h_{3,1})}\,(x_2; y_{3,1}) \\
&= \mathcal{I}_{(h_{1,3},h_{2,3},h_{3,2})}\,((x_1 \cdot x_2); y_{3,2}) + \mathcal{I}_{(h_{1,3},h_{2,3},h_{3,1})}\,((x_1 \cdot x_2); y_{3,1}).
\end{aligned}
\tag{1.30}
$$

In the following, we consider $\mathcal{I}_{(h_{1,3},h_{2,3},h_{3,2})}\,((x_1 \cdot x_2); y_{3,2})$ as an example, and $\mathcal{I}_{(h_{1,3},h_{2,3},h_{3,1})}\,((x_1 \cdot x_2); y_{3,1})$ can be calculated accordingly. It turns out that the distribution of $y_{3,2} = h_{3,2}x_3 + w_{3,2}$ is difficult to obtain due to the complicated distribution of v_3. Fortunately, v_3 given in Eq. (1.15) has exactly the same expression as the output LLR from a check node in the widely used sum-product decoding algorithm [17]. As it is shown in [17], a random variable with such an expression, although tends to be less like Gaussian, can be well approximated by the Gaussian distribution. Motivated by this result, we approximate v_3 as

$$
v_3 \approx v_3' \triangleq h_v\,(x_1 \cdot x_2) + w_v
\tag{1.31}
$$

where w_v is a Gaussian noise with the distribution $w_v \sim \mathcal{N}(0, 1)$ and h_v is a variable (real) which will be explained later. Then the forwarded signal can be approximated by

$$
x_3 \approx x_3' \triangleq \beta v_3' = \beta\,(h_v\,(x_1 \cdot x_2) + w_v) \qquad \text{with } E\left\{|x_3'|^2\right\} = 1
\tag{1.32}
$$

where $\beta = 1/\sqrt{h_v^2 + 1}$ is a normalizing factor in order to satisfy the power constraint $E\left\{|x_3'|^2\right\} = 1$. As shown in Step 1 and Fig. 1.4, v_3 can be seen as the output of a virtual channel whose input is $x_1 \cdot x_2$. The approximation in Eq. (1.31) further simplifies the virtual channel as a Gaussian channel with a real channel coefficient h_v and an additive Gaussian noise w_v, as shown in Fig. 1.6. Based on this approximation, we have

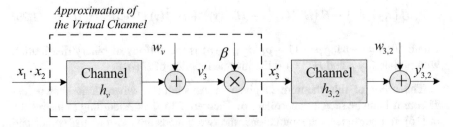

Fig. 1.6 Approximation of the virtual channel in AF-SoftNC

$$C_{\text{BPSK}}(h_v) \triangleq \mathcal{I}_{(h_v)}\left((x_1 \cdot x_2); x_3'\right) \approx \mathcal{I}_{(h_{1,3}, h_{2,3})}\left((x_1 \cdot x_2); x_3\right) \tag{1.33}$$

Since it is shown in *Step 1* that the bounds of $\mathcal{I}_{\text{step1}}$, which is given in Theorem 1, is very close to the true mutual information, we take the upper bound as the approximation of $\mathcal{I}_{\text{step1}}$. That is,

$$\mathcal{I}_{\text{step1}} \triangleq \mathcal{I}_{(h_{1,3}, h_{2,3})}\left((x_1 \cdot x_2); x_3\right) \approx \mathcal{G}\left(C_{\text{BPSK}}(h_{1,3}), C_{\text{BPSK}}(h_{2,3})\right). \tag{1.34}$$

By considering Eqs. (1.33) and (1.34) together, we have the channel coefficient h_v as

$$h_v = C_{\text{BPSK}}^{-1}\left(\mathcal{G}\left(C_{\text{BPSK}}(h_{1,3}), C_{\text{BPSK}}(h_{2,3})\right)\right) \tag{1.35}$$

where $C_{\text{BPSK}}^{-1}(\cdot)$ is the reverse function of $C_{\text{BPSK}}(\cdot)$, and its numerical calculation can be referred to [15]. According to Eq. (1.31) and as shown in Fig. 1.6, the received signal at node N_2 is then given by

$$y_{3,2} \approx y_{3,2}' = h_{3,2}x_3' + w_{3,2} = h_{3,2}\beta h_v(x_1 \cdot x_2) + h_{3,2}\beta w_v + w_{3,2}. \tag{1.36}$$

Given channel coefficients $h_{1,3}$ and $h_{3,2}$, we have

$$\begin{aligned} \mathcal{I}_{(h_{1,3}, h_{2,3}, h_{3,2})}(x_1; y_{3,2}) &= \mathcal{I}_{(h_{1,3}, h_{2,3}, h_{3,2})}\left((x_1 \cdot x_2); y_{3,2}\right) \\ &\approx \mathcal{I}_{(h_v, h_{3,2})}\left((x_1 \cdot x_2); y_{3,2}'\right) = C_{\text{BPSK}}\left(\sqrt{\frac{h_v^2 |h_{3,2}|^2}{h_v^2 + |h_{3,2}|^2 + 1}}\right). \end{aligned} \tag{1.37}$$

$\mathcal{I}_{(h_{1,3}, h_{2,3}, h_{3,1})}(x_2; y_{3,1})$ can be approximated by using the same procedure in the approximation of $\mathcal{I}_{(h_{1,3}, h_{2,3}, h_{3,2})}(x_1; y_{3,2})$. Finally by substituting these results into Eq. (1.30), we have

$$\mathcal{I}_{\text{step2}} \approx C_{\text{BPSK}}\left(\sqrt{\frac{h_v^2 |h_{3,2}|^2}{h_v^2 + |h_{3,2}|^2 + 1}}\right) + C_{\text{BPSK}}\left(\sqrt{\frac{h_v^2 |h_{3,1}|^2}{h_v^2 + |h_{3,1}|^2 + 1}}\right). \tag{1.38}$$

Again, based on the extensive simulations, it is found that the approximation in Eq. (1.38) is close to the true mutual information.

Step 3: *Approximation of $\mathcal{R}$*

As defined in Eq. (24), $\mathcal{R}$ is the maximum information that can be exchanged between the two end nodes through both the relay link and the direct link. Take node N_2. It receives $y_{1,2}$ from node N_1 (direct link) in the first time slot and $y_{3,2}$ from node N_3 (relay link) in the third time slot. Due to the fact that node N_2 knows its self-information x_2 and also based on the good approximation in Eq. (1.36), x_2 can be removed from $y_{3,2}$ by

$$\widetilde{y_{3,2}} \triangleq y_{3,2} \cdot x_2 \approx y_{3,2}' \cdot x_2 = h_{3,2}\beta h_v x_1 + h_{3,2}\beta w_v + w_{3,2}. \tag{1.39}$$

Assume that node N_2 has the channel state information of all the three channels $(h_{1,3}, h_{1,2}, \text{and } h_{3,2})$, it can perform MRC to $\tilde{y}_{3,2}$ and $y_{1,2}$, whose SNR values are $\left(h_v{}^2|h_{3,2}|^2\right) / \left(h_v^2 + |h_{3,2}|^2 + 1\right)$ and $|h_{1,2}|^2$, respectively. As a result, the final SNR of the combined signal can be approximated as $\left(h_v{}^2|h_{3,2}|^2\right) / \left(h_v^2 + |h_{3,2}|^2 + 1\right) + |h_{1,2}|^2$. Then, the total information of x_1 obtained at node N_2 through both the direct link and the relay link can be approximated as

$$
\begin{aligned}
\mathcal{I}_{\left(h_{1,2},h_{1,3},h_{2,3},h_{3,2}\right)} &(x_1; y_{3,2}, y_{1,2}) \approx \mathcal{I}_{\left(h_{1,2},h_{3,2},h_v\right)} (x_1; \tilde{y}_{3,2}, y_{1,2}) \\
&= C_{\text{BPSK}} \left(\sqrt{\frac{h_v{}^2|h_{3,2}|^2}{h_v^2 + |h_{3,2}|^2 + 1}} + |h_{1,2}|^2 \right).
\end{aligned}
\tag{1.40}
$$

By the same procedure, we can obtain the approximation of $\mathcal{I}_{\left(h_{1,2},h_{1,3},h_{2,3},h_{3,1}\right)}$ $(x_2; y_{3,1}, y_{2,1})$. By averaging over all channel fading coefficients, the ergodic mutual information rate of AF-SoftNC can be approximated by

$$
\begin{aligned}
&\mathcal{R}_{\text{AF-SoftNC}} \approx \\
&\underset{h_{i,j}}{E} \left\{ C_{\text{BPSK}} \left(\sqrt{\frac{h_v{}^2|h_{3,2}|^2}{h_v^2 + |h_{3,2}|^2 + 1}} + |h_{1,2}|^2 \right) + C_{\text{BPSK}} \left(\sqrt{\frac{h_v{}^2|h_{3,1}|^2}{h_v^2 + |h_{3,1}|^2 + 1}} + |h_{2,1}|^2 \right) \right\}.
\end{aligned}
\tag{1.41}
$$

1.5.2 Analysis of SBF-SoftNC

The analysis procedure for SBF-SoftNC is similar to that for AF-SoftNC protocol and is also taken in three steps. As shown in Theorem 1, the approximation of $\mathcal{I}_{\text{step1}}$ is the same in AF-SoftNC and SBF-SoftNC. We now discuss the following two steps in the analysis of SBF-SoftNC.

Step 2: Approximation of $\mathcal{I}_{\text{step2}}$

Since each end node knows its self-information, Eq. (1.30) still holds for SBF-SoftNC. As shown in Eq. (13), the soft bit generated at the relay node is in fact the "tanh" function of the LLR value. Since the LLR of $x_1 \cdot x_2$ can be approximated by $h_v (x_1 \cdot x_2) + w_v$ as in Eq. (1.31), it is reasonable to use a new random variable $\tanh (h_v (x_1 \cdot x_2) + w_v)$ to approximate the soft bit of $x_1 \cdot x_2$, i.e., v_3 in SBF-SoftNC. That is,

$$
v_3 \approx v_3'' \triangleq \tanh (h_v (x_1 \cdot x_2) + w_v)
\tag{1.42}
$$

where h_v and w_v have the same meaning as Eq. (1.31). Similar to Eq. (1.32), the forwarded signal can then be approximated by

$$x_3 \approx x_3'' \triangleq \theta \tanh\left(h_v\left(x_1 \cdot x_2\right) + w_v\right) \qquad \text{with } E\left\{\left|x_3''\right|^2\right\} = 1 \qquad (1.43)$$

where θ is a normalizing factor to satisfy the constraint $E\left\{\left|x_3''\right|^2\right\} = 1$. It can be seen from Eq. (1.42) that in the case of SBF-SoftNC the virtual channel in Fig. 1.4 can be approximated with the Gaussian channel followed by a soft bit transform ("tanh" function), as shown in Fig. 1.6. As a result, the received signal at the end node N_2, $y_{3,2}$, can be approximated by

$$y_{3,2} \approx y_{3,2}'' = h_{3,2}\theta \tanh\left(h_v\left(x_1 \cdot x_2\right) + w_v\right) + w_{3,2}. \qquad (1.44)$$

However, it is still difficult to calculate the mutual information between $x_1 \cdot x_2$ and $y_{3,2}''$, i.e., $\mathcal{I}\left(\left(x_1 \cdot x_2\right); y_{3,2}''\right)$. According to the conclusion of the capacity investigation of the memoryless relay protocols in OWRC [11], the AF protocol and the demodulate-and-forward (i.e., forward $\text{sign}(\hat{y}_{1,3}^{d}[m])$ in Eq. (1.11) to the destination node) protocol can approach the maximum mutual information performance (the mutual information of the SBF protocol) in low SNR region and high SNR region, respectively. By generalizing this result into the TWRC case, we further approximate $\mathcal{I}\left(\left(x_1 \cdot x_2\right); y_{3,2}''\right)$ by

$$\mathcal{I}\left(\left(x_1 \cdot x_2\right); y_{3,2}''\right) = \max\left\{\mathcal{I}_{\left(h_{3,2},h_v\right)}\left(\left(x_1 \cdot x_2\right); y_{3,2}'\right), \mathcal{I}_{\left(h_{3,2},h_v\right)}\left(\left(x_1 \cdot x_2\right); y_{3,2}^{\text{Hard}}\right)\right\} \qquad (1.45)$$

where $y_{3,2}^{\text{Hard}} \triangleq h_{3,2}\text{sign}\left(h_v\left(x_1 \cdot x_2\right) + w_v\right) + w_{3,2}$ and $\mathcal{I}_{\left(h_{3,2},h_v\right)}\left(\left(x_1 \cdot x_2\right); y_{3,2}'\right)$ is given in Eq. (1.37). The calculation of $\mathcal{I}_{\left(h_{3,2},h_v\right)}\left(\left(x_1 \cdot x_2\right); y_{3,2}^{\text{Hard}}\right)$ is given in Appendix 2. Finally, we obtain

$$\mathcal{I}_{\left(h_{1,3},h_{2,3},h_{3,2}\right)}\left(x_1; y_{3,2}\right) = \mathcal{I}_{\left(h_{1,3},h_{2,3},h_{3,2}\right)}\left(\left(x_1 \cdot x_2\right); y_{3,2}\right) \approx \mathcal{I}_{\left(h_v,h_{3,2}\right)}\left(\left(x_1 \cdot x_2\right); y_{3,2}''\right)$$

$$\approx \max\left\{C_{\text{BPSK}}\left(\sqrt{\frac{h_v^2\left|h_{3,2}\right|^2}{h_v^2 + \left|h_{3,2}\right|^2 + 1}}\right), \mathcal{G}\left(C_{\text{BPSK}}^{\text{Hard}}\left(h_v\right), C_{\text{BPSK}}\left(h_{3,2}\right)\right)\right\}. \qquad (1.46)$$

$\mathcal{I}_{\left(h_{1,3},h_{2,3},h_{3,1}\right)}\left(x_2; y_{3,1}\right)$ can be found in the same way. Finally, $\mathcal{I}_{\text{step2}}$ can be approximated by

$$\mathcal{I}_{\text{step2}} \approx \max\left\{C_{\text{BPSK}}\left(\sqrt{\frac{h_v^2\left|h_{3,2}\right|^2}{h_v^2 + \left|h_{3,2}\right|^2 + 1}}\right), \mathcal{G}\left(C_{\text{BPSK}}^{\text{Hard}}\left(h_v\right), C_{\text{BPSK}}\left(h_{3,2}\right)\right)\right\}$$

$$+ \max\left\{C_{\text{BPSK}}\left(\sqrt{\frac{h_v^2\left|h_{3,1}\right|^2}{h_v^2 + \left|h_{3,1}\right|^2 + 1}}\right), \mathcal{G}\left(C_{\text{BPSK}}^{\text{Hard}}\left(h_v\right), C_{\text{BPSK}}\left(h_{3,1}\right)\right)\right\}. \qquad (1.47)$$

Based on the extensive simulations, it is found that the approximation in Eq. (1.47) is close to the true mutual information.

Step 3: *Approximation of* $\mathcal{R}$

Since v_3 in Eq. (1.42) is the approximation of the soft bit of $x_1 \cdot x_2$, i.e., its MMSE estimation, it can also be regarded as $x_1 \cdot x_2$ plus certain noise. Then, similar to the approximation in Sect. 1.5.1, the node N_2 first removes its self-information from the received signal $y_{3,2}$ as follows:

$$\tilde{y}''_{3,2} = y_{3,2} \cdot x_2 \approx y''_{3,2} \cdot x_2 \approx h''_{v,2} x_1 + w''_{v,2} \tag{1.48}$$

where the real variable $h''_{v,2}$ is the virtual channel coefficient and $w''_{v,2}$ includes the noise $w_{3,2}$ introduced at N_2 and the equivalent noise introduced at the relay node and is approximated to be Gaussian distributed, i.e., $w''_{v,2} \sim \mathcal{N}(0,1)$. Then the virtual channel coefficient $h''_{v,2}$ can be determined by

$$
\begin{aligned}
h''_{v,2} &= C^{-1}_{\text{BPSK}} \left(\mathcal{I}_{h_{1,3},h_{2,3},h_{3,2}} (x_1 ; y_{3,2}) \right) \approx C^{-1}_{\text{BPSK}} \left(\mathcal{I}_{h_v,h_{3,2}} \left((x_1 \cdot x_2) ; y''_{3,2} \right) \right) \\
&\approx \max \left\{ \sqrt{\frac{h_v{}^2 |h_{3,2}|^2}{h_v{}^2 + |h_{3,2}|^2 + 1}}, C^{-1}_{\text{BPSK}} \left(\mathcal{G} \left(C^{\text{Hard}}_{\text{BPSK}} (h_v), C_{\text{BPSK}} (h_{3,2}) \right) \right) \right\}.
\end{aligned}
\tag{1.49}
$$

After combining $\tilde{y}''_{3,2}$ and $y_{1,2}$, the SNR of the new signal is $\left(h''_{v,2} \right)^2 + |h_{1,2}|^2$. The total information of x_1 obtained at node N_2 through both the direct link and the relay link can be approximated as

$$\mathcal{I}(x_1 ; y_{3,2}, y_{1,2}) \approx \mathcal{I}\left(x_1 ; \tilde{y}''_{3,2}, y_{1,2} \right) \approx C_{\text{BPSK}} \left(\sqrt{\left(h''_{v,2} \right)^2 + |h_{1,2}|^2} \right). \tag{1.50}$$

Note that in Eq. (1.48) we assume that $w''_{v,2} \sim \mathcal{N}(0,1)$ in order to combine $\tilde{y}''_{3,2}$ and $y_{1,2}$ with MRC. We argue that different assumptions on the distribution of $w''_{v,2}$ do not affect the final result in Eq. (1.50) too much because, based on the conclusion in [16], the mutual information between x_1 and the two binary input symmetric memoryless output (BISMO) channels' outputs, $\tilde{y}''_{3,2}$ and $y_{1,2}$, does not depend too much on the distribution of each channel when $\mathcal{I}(x_1 ; y_{1,2})$ and $\mathcal{I}\left(x_1 ; \tilde{y}''_{3,2} \right)$ are given.

In a similar way, we can obtain the approximation $\mathcal{I}(x_2 ; y_{3,1}, y_{2,1}) \approx C_{\text{BPSK}} \left(\sqrt{\left(h''_{v,1} \right)^2 + |h_{2,1}|^2} \right)$ where the virtual channel coefficient $h''_{v,1}$ is similar to Eq. (1.49)

$$
\begin{aligned}
h''_{v,1} &= C^{-1}_{\text{BPSK}} \left(\mathcal{I}_{h_{1,3},h_{2,3},h_{3,1}} (x_2 ; y_{3,1}) \right) \approx C^{-1}_{\text{BPSK}} \left(\mathcal{I}_{h_v,h_{3,1}} \left((x_1 \cdot x_2) ; y''_{3,1} \right) \right) \\
&\approx \max \left\{ \sqrt{\frac{h_v{}^2 |h_{3,1}|^2}{h_v{}^2 + |h_{3,1}|^2 + 1}}, C^{-1}_{\text{BPSK}} \left(\mathcal{G} \left(C^{\text{Hard}}_{\text{BPSK}} (h_v), C_{\text{BPSK}} (h_{3,1}) \right) \right) \right\}.
\end{aligned}
\tag{1.51}
$$

By averaging over all the channel realizations, we can obtain the approximation of $\mathcal{R}$ for the SBF-SoftNC as

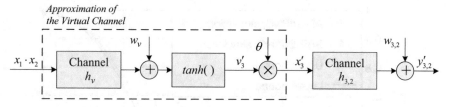

Fig. 1.7 Approximation of the virtual channel in SBF-SoftNC

$$\mathcal{R}_{\text{SBF-SoftNC}} \approx \underset{h_{i,j}}{E} \left\{ C_{\text{BPSK}} \left(\sqrt{\left(h_{v,2}'' \right)^2 + |h_{1,2}|^2} \right) \right\} + \underset{h_{i,j}}{E} \left\{ C_{\text{BPSK}} \left(\sqrt{\left(h_{v,1}'' \right)^2 + |h_{2,1}|^2} \right) \right\}.$$
(1.52)

1.5.3   Numerical Simulation

In this part, we first present simulation results to show the accuracy of the bounds and approximations which have been derived in the above parts, in the first and second steps of the MAIR analysis. We then compare the MAIR performance of the proposed AF-SoftNC and SBF-SoftNC with that of the conventional relay protocols in TWRC.

1.5.3.1   Simulation Results of Bounds and Approximations

We first investigate the accuracy of the upper and lower bounds for $\mathcal{I}_{\text{step1}}$, given in Eq. (1.28) in Theorem 1. The performance of the theoretical bounds and that of the true mutual information in AF-SoftNC and SBF-SoftNC designs are compared in Fig. 1.7 with different channel gains $\gamma_{1,3} \triangleq |h_{1,3}|^2$ and $\gamma_{2,3} \triangleq |h_{2,3}|^2$. In the simulation of this subsection, we assume $\gamma_{i,j} = \gamma_{j,i}$ $\forall i,j \in \{1, 2, 3\}$. Two groups of curves are presented, where $\gamma_{1,3}$ is set to 0 dB for one group and is set to be the same as $\gamma_{2,3}$ for the other group. The true mutual information of $\mathcal{I}_{\text{step1}}$ for AF-SoftNC and SBF-SoftNC is calculated according to its definition. That is,

$$\mathcal{I}\left((x_1 \cdot x_2) ; x_3 \right) = \mathcal{H}(x_3) - \mathcal{H}(x_3 | (x_1 \cdot x_2))$$
$$= \mathcal{H}(x_3) - \mathcal{H}(x_3 | (x_1 = x_2 = \pm 1)).$$
(1.53)

As shown in Fig. 1.8, both the upper and the lower bounds are close to $\mathcal{I}_{\text{step1}}$ and can be seen as good approximations in the analysis of AF-SoftNC and SBF-SoftNC systems.

Next, in Fig. 1.9, we present some simulation results in the approximation of $\mathcal{I}_{\text{step2}}$ with different $\gamma_{1,3}$ and $\gamma_{2,3}$. In the simulation, $\gamma_{2,3} = \gamma_{3,2}$ is set to 0 dB for one group of curves and is set to the same as $\gamma_{1,3} = \gamma_{3,1} = \gamma_{3,2} = \gamma_{2,3}$

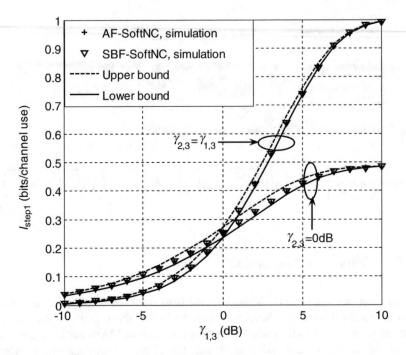

Fig. 1.8 Bounds of $\mathcal{I}_{\text{step1}}$ for AF-SoftNC and SBF-SoftNC

for the other group. The curves of the approximation results for AF-SoftNC and SBF-SoftNC are obtained according to Eqs. (1.38) and (1.47), respectively. The true mutual information is given by the definition

$$
\begin{aligned}
\mathcal{I}(x_1; y_{3,2}) + \mathcal{I}(x_2; y_{3,1}) &= \mathcal{I}((x_1 \cdot x_2); y_{3,2}) + \mathcal{I}((x_1 \cdot x_2); y_{3,1}) \\
&= \mathcal{H}(y_{3,2}) - \mathcal{H}(y_{3,2} \,|\, x_1 = x_2) + \mathcal{H}(y_{3,1}) - \mathcal{H}(y_{3,1} \,|\, x_1 = x_2).
\end{aligned}
\tag{1.54}
$$

where $\mathcal{H}(x)$ denotes the entropy of a random variable x.

It is shown in Fig. 1.9 that the proposed approximation results are close to $\mathcal{I}_{\text{step2}}$ for almost the whole range of SNR values. It is also shown that the SBF-SoftNC has a better performance than the AF-SoftNC design especially when the SNR is high, which has also been shown in Eq. (1.47). This result is consistent with that in the comparison of AF and SBF in OWRC [11].

1.5.3.2 MAIR Performance Comparison

In this part, we compare the MAIR performance of AF-SoftNC and SBF-SoftNC with that of three conventional two-way relay protocols. The first conventional protocol is amplify and forward without network coding, referred to as AF-NNC,

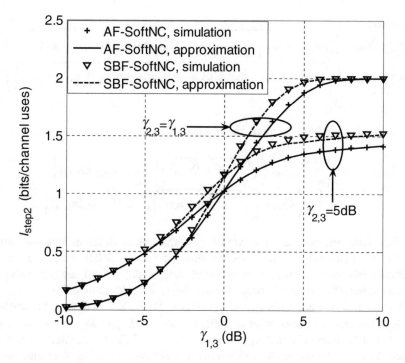

Fig. 1.9 Approximation of $\mathcal{I}_{\text{step2}}$ for AF-SoftNC and SBF-SoftNC

where the relay node simply amplifies and forwards the two received packets, respectively, in the third and the fourth time slots. The second conventional protocol, referred to as SBF-NNC, is the same as AF-NNC except that the relay node simply forwards the soft bit of the two received packets rather than the amplified packets in the third and the fourth time slots. The third one is the SNC protocol similar to [4], where the relay node decodes the two received packets from the two end nodes separately in an explicit manner and then combines the packets into a network-coded packet for forwarding. For the AF-NNC protocol, the MAIR is given by

$$
\begin{aligned}
\mathcal{R}_{\text{AF-NNC}} = {} & \tfrac{3}{4} \underset{h_{i,j}}{E} \left\{ C_{\text{BPSK}} \left(\sqrt{|h_{2,1}|^2 + \frac{|h_{2,3}|^2 \cdot |h_{3,1}|^2}{|h_{2,3}|^2 + |h_{3,1}|^2 + 1}} \right) \right\} \\
& + \tfrac{3}{4} \underset{h_{i,j}}{E} \left\{ C_{\text{BPSK}} \left(\sqrt{|h_{1,2}|^2 + \frac{|h_{1,3}|^2 \cdot |h_{3,2}|^2}{|h_{1,3}|^2 + |h_{3,2}|^2 + 1}} \right) \right\} .
\end{aligned}
\tag{1.55}
$$

where the coefficient 3/4 is added since 4 time slots are needed in the whole relay process. For SBF-NNC, the MAIR is derived in a similar way as in Eq. (1.45)

$\mathcal{R}_{\text{SBF-NNC}}$

$$\approx \frac{3}{4} \underset{h_{i,j}}{E} \left\{ C_{\text{BPSK}} \left(\sqrt{|h_{2,1}|^2 + \max \left\{ \frac{|h_{2,3}|^2 \cdot |h_{3,1}|^2}{|h_{2,3}|^2 + |h_{3,1}|^2 + 1}, C_{\text{BPSK}}^{-1} \left(\mathcal{G} \left(C_{\text{BPSK}}^{\text{Hard}} (h_{2,3}), C_{\text{BPSK}} (h_{3,1}) \right) \right) \right\}} \right) \right\}$$
$$+ \frac{3}{4} \underset{h_{i,j}}{E} \left\{ C_{\text{BPSK}} \left(\sqrt{|h_{1,2}|^2 + \max \left\{ \frac{|h_{1,3}|^2 \cdot |h_{3,2}|^2}{|h_{1,3}|^2 + |h_{3,2}|^2 + 1}, C_{\text{BPSK}}^{-1} \left(\mathcal{G} \left(C_{\text{BPSK}}^{\text{Hard}} (h_{1,3}), C_{\text{BPSK}} (h_{3,2}) \right) \right) \right\}} \right) \right\}$$
$$(1.56)$$

For the traditional SNC protocol, the MAIR is given by

$$\mathcal{R}_{\text{SNC}} = \min \left\{ \underset{h_{i,j}}{E} \left\{ C_{\text{BPSK}} \left(\sqrt{|h_{1,2}|^2 + |h_{3,2}|^2} \right) \right\}, \underset{h_{i,j}}{E} \left\{ C_{\text{BPSK}} (h_{1,3}) \right\} \right\}$$
$$+ \min \left\{ \underset{h_{i,j}}{E} \left\{ C_{\text{BPSK}} \left(\sqrt{|h_{2,1}|^2 + |h_{3,1}|^2} \right) \right\}, \underset{h_{i,j}}{E} \left\{ C_{\text{BPSK}} (h_{2,3}) \right\} \right\} . \quad (1.57)$$

In Fig. 1.10, we compare the MAIR performance of different two-way relay protocols when fixing the qualities (in terms of average SNR) of relay channels (the channels between the end nodes and the relay node) and varying the quality of direct channel (the channel between the two end nodes). We assume Rayleigh fading and symmetric average SNR $\lambda_{i,j} = \lambda_{j,i}$ $\forall i,j \in \{1,2,3\}$. Specifically, the average SNR of the relay channels is set to 5 dB, i.e., $\lambda_{3,1} = \lambda_{3,2} = \lambda_{1,3} = \lambda_{2,3} = \sqrt{10}$, and that of the direct channel $\lambda_{1,2}$ varies from -5 to 15 dB. In our simulation, "one channel use" means one transmission process of SoftNC (including three time slots). As shown Fig. 1.10, SoftNC schemes always outperform traditional nonnetwork coding schemes. When the SNR of the direct channel is larger than 2 dB, SoftNC schemes outperform the traditional SNC scheme even though the latter involves more processing in terms of channel decoding and re-encoding at the relay. Note that the MAIR of the traditional SNC scheme is constant because it is constrained by the unchanged relay links.

In Fig. 1.11, we investigate the effect of the relay channel quality on the MAIR performance of different two-way relay protocols. Specifically, in the simulation, the average channel SNR $\lambda_{2,3} = \lambda_{3,2} = \lambda_{2,1} = \lambda_{1,2} = 0$ dB and the average SNR of the relay channel between N_1 and N_3, $\lambda_{1,3} = \lambda_{3,1} = -5$ to 15dB. As shown in Fig. 1.11, SoftNC schemes also always outperform traditional nonnetwork coding schemes. It is better than the traditional SNC scheme except when $\gamma_{1,3} = \gamma_{3,1}$ falls in the range of 2–7 dB. This is because the transmission rate of SNC is severely constrained by $\gamma_{1,3}$ when it is much less $\gamma_{2,3}$, and the transmission rate of SNC is constrained to constant by $\gamma_{2,3}$ when $\gamma_{1,3}$ is larger than 5 dB.

From the simulation, we can see that when compared to the schemes with the same complexity (no channel coding used at the relay), our proposed SoftNC schemes are always better than the schemes without network coding. Even compared to the traditional SNC scheme, which is much more complex than our schemes due to the channel decoding and re-encoding at the lay, SoftNC still outperforms in some scenarios.

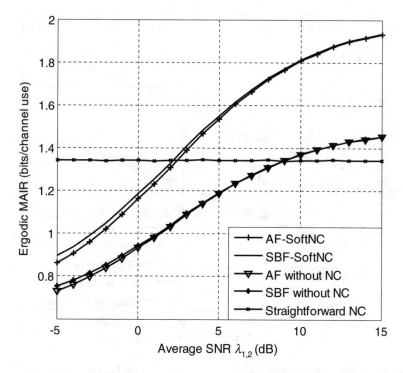

Fig. 1.10 Performance comparison when $\lambda_{1,3} = \lambda_{2,3} = 5$ dB and $\lambda_{1,2}$ varies from -5 to 15 dB

1.6   Summary

In this chapter, we propose two new network coding schemes for wireless TWRC, AF-SoftNC, and SBF-SoftNC. In both schemes, the relay forwards "soft decision" information on the XOR of the two bits, received from the two end nodes through orthogonal channels. In AF-SoftNC, the log-likelihood ratio is forwarded; in SBF-SoftNC, the MMSE estimate is forwarded. In this way, only symbol-level operation, rather than packet-level channel decoding and re-encoding operation (e.g., the traditional network coding), is required at the relay.

We analyze the ergodic MAIR of the SoftNC schemes. Compared to traditional SNC scheme, our proposed SoftNC schemes perform better under some common scenarios, such as the case when the direct link has a better quality, although the relay is appreciably less complex. For implementation, in time-varying wireless networks, the relay node may adaptively select SoftNC or traditional SNC based on the channel quality and on its own capability (energy, hardware, and so on).

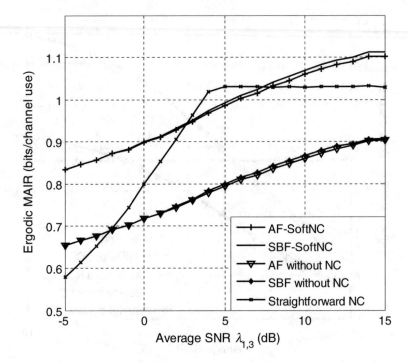

Fig. 1.11 Performance comparison when $\lambda_{1,2} = \lambda_{3,2} = 0$ dB and $\lambda_{1,3}$ varies from -5 to 15 dB

Appendix 1: Proof of Theorem 1

Note that the multiplication of two BPSK symbols in the transmitted packet, $x_1 \cdot x_2$, is equivalent to the XOR of two binary bits in the corresponding code words, $d_1 \oplus d_2$. Since Gaussian channel belongs to BISMO channels, the following inequalities can be obtained from Theorem 2 in [16] directly

$$
\begin{aligned}
C_{\text{bpsk}}\left(h_{1,3}\right) C_{\text{bpsk}}\left(h_{2,3}\right) &\leq \mathcal{I}\left(\left(x_1 \cdot x_2\right); y_{1,3}, y_{2,3}\right) \\
&= \mathcal{I}\left(\left(x_1 \cdot x_2\right); v_{1,3}, v_{2,3}\right) \leq \mathcal{G}\left(C_{\text{bpsk}}\left(h_{1,3}\right), C_{\text{bpsk}}\left(h_{2,3}\right)\right).
\end{aligned}
\tag{1.58}
$$

According to the signal processing Eq. (1.15) in AF-SoftNC and that Eq. (1.17) in SBF-SoftNC, it can be found that no information about $x_1 \cdot x_2$ is lost, i.e., $\mathcal{I}\left(\left(x_1 \cdot x_2\right); y_{1,3}, y_{2,3}\right) = \mathcal{I}\left(\left(x_1 \cdot x_2\right); x_3\right)$. We take the SBF-SoftNC protocol as an example to prove it

$$
\begin{aligned}
\mathcal{I}\left((x_1 \cdot x_2); y_{1,3}, y_{2,3}\right) &= \mathcal{H}\left(x_1 \cdot x_2\right) - \mathcal{H}\left((x_1 \cdot x_2) \mid y_{1,3}, y_{2,3}\right) \\
&= 1 - H\left(\Pr\left((x_1 \cdot x_2) = -1 \mid y_{1,3}, y_{2,3}\right)\right) \\
&= 1 - H\left(\frac{1 - \left(\Pr((x_1 \cdot x_2)=1 \mid y_{1,3}, y_{2,3})\right) - \Pr((x_1 \cdot x_2)=-1 \mid y_{1,3}, y_{2,3})}{2}\right). \\
&= 1 - H\left(\frac{1 - x_3}{2}\right) = 1 - H\left(\Pr\left((x_1 \cdot x_2) = -1 \mid x_3\right)\right) \\
&= 1 - \mathcal{H}\left((x_1 \cdot x_2) \mid x_3\right) = \mathcal{I}\left((x_1 \cdot x_2); x_3\right)
\end{aligned}
\tag{1.59}
$$

This completes the proof.

Appendix 2: Mutual Information for Demodulate and Forward

This appendix shows the approximation of $\mathcal{I}_{(h_{3,2}, h_v)}\left((x_1 \cdot x_2); y_{3,2}^{\mathrm{Hard}}\right)$ in Eq. (1.45), which can be seen as the mutual information between the source and destination of a two-hop memoryless relay channel, where $x_1 \cdot x_2$ is the transmitted signal from the source; h_v and $h_{3,2}$ are the channel coefficients of the two hops, respectively; and w_v and $w_{3,2}$ are the noises at relay node and destination node, respectively. Since it is assumed that the demodulate-and-forward protocol is used at the relay, the mutual information of the first hop with hard decision is given by

$$
\begin{aligned}
&\mathcal{I}\left((x_1 \cdot x_2); \operatorname{sign}\left(h_v\left(x_1 \cdot x_2\right) + w_v\right)\right) \\
&= 1 - H\left(\Pr\left((x_1 \cdot x_2) \neq \operatorname{sign}\left(h_v\left(x_1 \cdot x_2\right) + w_v\right)\right)\right) \\
&= 1 - H\left(1 - \Phi_{0,1}\left(\left|h_v\right|\right)\right) \triangleq C_{\mathrm{BPSK}}^{\mathrm{Hard}}\left(h_v\right).
\end{aligned}
\tag{1.60}
$$

where $\Phi_{0,1}(\cdot)$ is the cumulative distribution function of a normal distribution $\mathcal{N}(0,1)$. It can be seen that the capacity of the second hop is $C_{\mathrm{BPSK}}(h_{3,2})$. Although it is difficult to calculate the exact formula of $\mathcal{I}_{(h_{3,2}, h_v)}\left((x_1 \cdot x_2); y_{3,2}^{\mathrm{Hard}}\right)$, the following theorem provides a tight upper bound and a tight lower bound.

Lemma 2. For a serial concatenated channel which consists of two binary symmetric channels (BSCs) with the crossover probabilities being p_1 and p_2, and the mutual information being $I_1 = H(p_1)$ and $I_2 = H(p_2)$, respectively, as shown in Fig. 1.12. The mutual information of the serial concatenated channel is

$$
I_{\mathrm{SC}} = \mathcal{G}\left(I_1, I_2\right).
\tag{1.61}
$$

Proof. Since the crossover probabilities of the two BSC are p_1 and p_2, respectively, the crossover probability of the serial concatenated channel is

$$
p = p_1\left(1 - p_2\right) + p_2\left(1 - p_1\right) = p_1 + p_2 - 2 p_1 p_2.
\tag{1.62}
$$

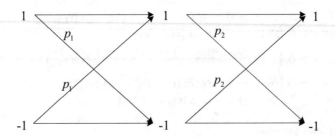

Fig. 1.12 The serial concatenation of two BSCs

According to the definition of $H(p)$ in Eq. (1.29), the mutual information of the serial concatenated channel is

$$I_{SC} = 1 - H(p) = \mathcal{G}(I_1, I_2). \tag{1.63}$$

Theorem 3. Considering two-hop memoryless relay channels, denoted by $S \rightarrow R \rightarrow T$, the channel in the first hop is a BSC with mutual information I_{hop1} and that in the second hop is a binary input symmetric memoryless output (BISMO) channel with mutual information I_{hop2}. Then the capacity between S and T, denoted by I_{total}, is bounded by

$$I_{hop1} \cdot I_{hop2} \leq I_{total} \leq \mathcal{G}(I_{hop1}, I_{hop2}). \tag{1.64}$$

Proof. As shown in [16], the BISMO channel in the second hop can be decomposed into a set of BSCs. Define $J \triangleq \{j_1, j_2, j_3, j_4, \cdots\}$ with $j_n \geq 0$ being the set which consists of the amplitude values of the possible outputs of the second channel. Figure 1.13 provides an example of the BISMO channel decomposition. Define the crossover probability of each BSC in the decomposition of the BISMO channel as

$$q(j_n) \triangleq \begin{cases} \Pr(T = -j_n \,|\, R = +1 \text{ and } T = \pm j_n) & \text{if } j_n \neq 0 \\ 1/2 & \text{if } j_n = 0 \end{cases}. \tag{1.65}$$

The mutual information of each BSC is given by

$$I(j_n) \triangleq \mathcal{I}(R; T \,|\, T = \pm j_n) = 1 - H(q(j_n)). \tag{1.66}$$

By averaging over all possible j_n, the mutual information of the BISMO channel in the second hop is given by

$$I_{hop2} = \underset{j_n \in J}{E} \{I(j_n)\}. \tag{1.67}$$

By considering the BSC channel in the first hop and the decomposition of the BISMO channel in the second hop together, the total serial concatenated channel

Fig. A.13 Decomposition a BISMO channel into a set of BSCs

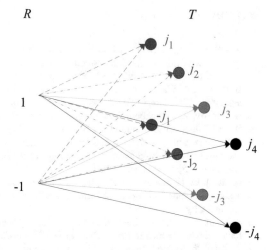

can be decomposed into a group of small serial concatenated channels with each being made up of two BSCs. According to Lemma 2, the mutual information of the small serial concatenated channel with the output j_n is given by $\mathcal{G}\left(I_{\text{hop1}}, I\left(j_n\right)\right)$. Averaging it over all possible j_n yields

$$I_{\text{total}} = \underset{j_n \in J}{E}\left\{\mathcal{G}\left(I_{\text{hop1}}, I\left(j_n\right)\right)\right\}. \tag{1.68}$$

Since it has been shown in [16] that $\mathcal{G}(a, b)$ is concave for both a and b and $\mathcal{G}(a, b) \geq a \cdot b$, it can be found that

$$I_{\text{hop1}} \cdot I_{\text{hop2}} = \underset{j_n \in J}{E}\left\{I_{\text{hop1}} I\left(j_n\right)\right\} \leq I_{\text{total}} \leq \mathcal{G}\left(I_{\text{hop1}}, \underset{j_n \in J}{E}\left\{I\left(j_n\right)\right\}\right) = \mathcal{G}\left(I_{\text{hop1}}, I_{\text{hop2}}\right). \tag{1.69}$$

The received signal $y_{3,2}^{\text{Hard}} \triangleq h_{3,2}\text{sign}\left(h_v\left(x_1 \cdot x_2\right) + w_v\right) + w_{3,2}$ can be regarded as the output of a serial concatenated channel. The channel in the first hop whose input is $x_1 \cdot x_2$ and the output is $\text{sign}\left(h_v\left(x_1 \cdot x_2\right) + w_v\right)$ can be regarded as a BSC. The Gaussian channel in the second hop belongs to BISMO channels, and it can be obtained directly from Theorem 3 that

$$C_{\text{BPSK}}^{\text{Hard}}\left(h_v\right) \cdot C_{\text{BPSK}}\left(h_{3,2}\right) \leq \mathcal{I}_{\left(h_{3,2}, h_v\right)}\left(\left(x_1 \cdot x_2\right); y_{3,2}^{\text{Hard}}\right) \leq \mathcal{G}\left(C_{\text{BPSK}}^{\text{Hard}}\left(h_v\right), C_{\text{BPSK}}\left(h_{3,2}\right)\right) \tag{1.70}$$

where $C_{\text{BPSK}}^{\text{Hard}}\left(h_v\right)$ and $C_{\text{BPSK}}(h_{3,2})$ denote the mutual information of the channels in the two hops, respectively.

References

1. S. Zhang, S. C. Liew, P. P. Lam, Hot topic: physical layer network coding, in *Proceedings of the MobiCom'06: the 12th Annual International Conference on Mobile Computing and Networking*, New York, NY, USA, 2006, pp. 358–365
2. B. Rankov, A. Wittneben, Achievable rate regions for the two-way relay channel, in *Proceedings of the IEEE International Symposium on Information Theory*, Seattle, WA, 2006
3. B. Rankov, A. Wittneben, Spectral efficient protocols for half duplex fading relay channels. IEEE J Select Areas Commun **25**(2), 379–389 (2007)
4. Y. Wu, P. A. Chou, S. Y. Kung, Information exchange in wireless networks with network coding and physical-layer broadcast, in *Proceedings of the 39th Annual Conference on Information Science and Systems (CISS)*, 2005
5. M. Yu, J. Li, H. Sadjadpour, Amplify-forward and decode-forward: the impact of location and capacity contour, in *Proceedings of IEEE Military Communications Conference (MILCOM)*, Atlantic City, NJ, October 2005
6. S. Zhang, Y. Zhu, S.-C. Liew, K. B. Letaief, Joint design of network coding and channel coding for wireless networks, in *Proceedings of IEEE WCNC2007*, Hong Kong, 2007
7. S. Yang, R. Koetter, Network coding over a noisy relay: a belief propagation approach, in *Proceedings of IEEE International Symposium on Information Theory (ISIT'07)*, Nice, France, 24–29 July 2007
8. W. Pu, C. Luo, S. Li, C. Chen, "Continuous network coding in wireless relay networks," in *Proceedings of the 27-th IEEE INFOCOM*, Phoenix, USA, April 2008
9. J. Hagenauer, Forward error correction for CDMA system, in *Proceedings of the IEEE Fourth International Symposium on Spread Spectrum Techniques and Application (ISSSTA'96)*, September 1996, pp. 566–569
10. K.S. Gomadam, S.A. Jafar, Optimal relay functionality for SNR maximization in memoryless relay networks. IEEE J Select Areas Commun **27**(2), 390–401 (2007)
11. K. S. Gomadam, S. A. Jafar, On the capacity of memoryless relay networks, *in Proceedings of IEEE ICC '06*, June 2006, pp. 1580–1585
12. J. Chen, M.P.C. Fossorier, Near optimal universal belief propagation based decoding of Low Density Parity Check codes. IEEE Trans Commun **50**(3), 406–414 (2002)
13. S.J. MacMullan, O.M. Collins, The capacity of orthogonal and bi-orthogonal codes on the Gaussian channel. IEEE Trans Inf Theory **44**(3), 1217–1322 (1998)
14. D. Tse, P. Viswanath, *Fundamentals of Wireless Communication* (Cambridge University Press, Cambridge, 2005)
15. S. ten Brink, G. Kramer, A. Ashikhmin, Design of low-density parity-check codes for modulation and detection. IEEE Trans Commun **52**, 670–678 (2004)
16. I. Land, S. Huettinger, P.A. Hoeher, J.B. Huber, Bounds on information combining. IEEE Trans Inf Theory **51**(2), 612–619 (2005)
17. S.Y. Chung, T.J. Richardson, R.L. Urbanke, Analysis of sum-product decoding of Low Density Parity Check codes using a Gaussian approximation. IEEE Trans Inf Theory **47**(2), 657–670 (2001)
18. S.-Y.R. Li, R.W. Yeung, N. Cai, Linear network coding. IEEE Trans Inf Theory **49**(2), 371–381 (2003)
19. F. Xue, S. Sandhu, PHY-layer network coding for broadcast channel with side information, in *Proceedings of IEEEInformation Theory Workshop, 2007, ITW '07,* September 2007.

Chapter 2
Throughput of Network Coding Nodes Employing Go-Back-N or Selective-Repeat Automatic Repeat ReQuest

Yang Qin and Lie-Liang Yang

Abstract The steady-state performance of general network coding nodes is investigated, when data are transmitted in packets based on the go-back-N automatic repeat request (GBN-ARQ) or selective-repeat ARQ (SR-ARQ) error-control scheme. A general network coding node is assumed to have H incoming links that provide packets for forming the coded packets transmitted by one outgoing link. Each of the incoming and outgoing links is assumed to have some buffers for temporarily storing the data packets. The state transitions of network coding nodes employing GBN-ARQ or SR-ARQ are analyzed, which shows that the operations of a general network coding node can be modeled by a finite state machine. Therefore, the expressions for the steady-state throughput of general network coding nodes are derived based on the properties of finite-state machines. Furthermore, the throughput performance of network coding nodes is investigated by both simulations and evaluation of the expressions obtained. The studies show that the simulation results converge closely to the numerical results, which justify the effectiveness of the analytical expressions derived. Furthermore, the studies show that the packet error rate, the capacity of buffer, and the number of incoming links may impose significant impact on the performance of general network coding nodes.

The original version of this chapter was revised. An erratum to this chapter can be found at DOI 10.1007/978-3-319-29770-5_6

Y. Qin (✉) • L.-L. Yang
School of Electronics and Computer Science, University of Southampton,
Southampton SO17 1BJ, UK
e-mail: yqinphd@gmail.com; lly@ecs.soton.ac.uk
http://www-mobile.ecs.soton.ac.uk

2.1   Introduction

Network coding deals with the problems of coding over packet networks. It has been recognized that the network coding assisted routing has the potential to outperform the conventional routing [1, 2]. Owing to the promised performance potential, various network protocols, such as the peer-to-peer content distribution protocols, based on the network coding principles have been proposed, as seen in [3, 4], for example. In literature, performance of communication systems employing network coding has been investigated, typically, under the assumption that data packets are transmitted reliably from one node to another without error [5]. However, in practical wired or wireless communication networks, the communication channels are always non-ideal and transmission errors may occur. Hence, error-control techniques are usually required in order to ensure high-reliability communications [2, 6].

In literature, network coding with feedback has first been studied in [7], which shows that using feedback may be beneficial to parameter adaptation, reliability enhancement, and packet acknowledgment in network coding systems. In [8, 9], the so-called *drop-when-seen* Automatic Retransmission reQuest (ARQ)-assisted network coding scheme has been proposed and investigated, in order to attain relative short transmission queues and to reduce the average decoding delay. In [10], a random network coding framework employing hybrid ARQ scheme has been proposed for real-time media broadcast over single-hop wireless networks. Furthermore, in order to overcome the problem that packets required for encoding a packet may arrive at the coding node asynchronously, the authors in [11] have proposed and investigated a network coding scheme, where a buffer is introduced to each of the incoming links of the coding node. Additionally, there are several network coding schemes having been proposed in order to improve the delay performance of network coding assisted networks. For example, [12] suggests to transmit uncoded packets at critical moments, so as to meet the delay requirement. By contrast, in [13], a feedback-based adaptive broadcast coding scheme has been developed, in order to reduce the delay for the applications where packets have to be accepted in order. Furthermore, some ARQ-assisted network coding schemes have been proposed and analyzed for single-hop networks. For example, as above-mentioned, in [10], a random network coding framework employing hybrid ARQ scheme has been proposed for single-hop real-time media broadcast. In [14], a queuing-based dynamic network coding scheme has been studied in the context of the single-hop networks consisting of one source node and multiple receivers.

The throughput and delay performance of some network coding nodes and Butterfly network have been investigated in [15–21], when various ARQ schemes are considered. Specifically, in [15], the maximum achievable throughput and the steady-state throughput of the Butterfly networks have been investigated and compared in the context of three types of ARQ schemes, namely, Stop-and-Wait (SW-ARQ), Go-Back-N (GBN-ARQ), and the Selective-Repeat (SR-ARQ). Correspondingly, the average burst delay and the standard deviation (SD) of burst delay of the Butterfly networks have been investigated and compared in [16], also when the SW-ARQ, GBN-ARQ, and the SR-ARQ are considered. Whereas, the references of [17–21] have all been contributed to the SW-ARQ, and investigated the performance of the network coding nodes and Butterfly networks from different perspectives, including throughput, delay, blocking probability, distribution of contents, etc. The studies show that, while the SW-ARQ has the advantage of simplicity for operation, it also has some shortcomings for operation in some communication scenarios. The deficiency of the SW-ARQ is that, after transmission of a packet, it has to stop transmission until receiving an ACKnowledgement (ACK) or a Negative ACKnowledgement (NACK) from the receiver. As shown in [15, 16], in comparison to the SR-ARQ, the GBN-ARQ, and SR-ARQ schemes are usually capable of providing better performance than the SW-ARQ scheme, while requiring relative higher implementation complexity than the SW-ARQ scheme. Therefore, in this chapter, we will study the steady-state throughput of the general network coding nodes, when communications between two nodes are based on the GBN-ARQ or the SR-ARQ.

Specifically, when the GBN-ARQ is studied, we assume that each link is associated with a buffer of arbitrary size. In this case, the transmitter can transmit all the packets in the transmission window without stopping to wait for the acknowledgement for the outstanding packets. However, when a NACK is received by the transmitter, the corrupted packet and the following packets are then all retransmitted. By contrast, when the SR-ARQ is employed, the transmitter keeps transmitting packets, until receiving a NACK. In this event, only the erroneous packet is retransmitted. In this chapter, the steady-state throughput of the general H-Inputs-Single-Output (HISO) network coding node is analyzed. The operations, properties, and transition probability of a HISO network coding node supported by the GBN-ARQ or SR-ARQ are analyzed in detail. Furthermore, by modeling the operations of network coding nodes as a finite-state machine (FSM) working in the principles of discrete-time Markov chain, we derive the steady-state throughput of the HISO network coding nodes. Finally, the throughput performance of network coding nodes employing the GBN-ARQ or SR-ARQ with different numbers of input links is investigated by both simulation and numerical approaches. Our performance results demonstrate that the simulation results converge well to the numerical results, which justify the effectiveness of our analytical expressions derived.

This chapter is organized as follows. In Sect. 2.2, we describe the system model for a network coding node, as well as the main assumptions and operational principles. In Sects. 2.3 and 2.4, we derive the steady-state throughput of the network coding node employing the GBN-ARQ and the SR-ARQ, respectively. Section 2.5 demonstrates the throughput performance. The final section of Sect. 2.6 summarizes the main findings.

2.2   System Models, Assumptions, and Transmission Operations

2.2.1   System Models

The system model considered for the GBN-ARQ is shown in Fig. 2.1a, which is constituted by three nodes, A, B, and C. Node A is an *abstract* source node with

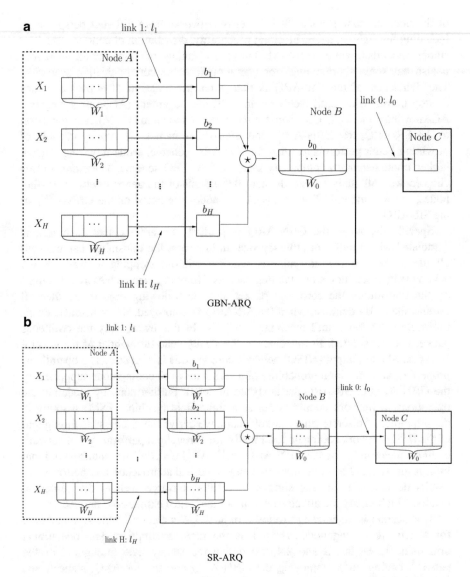

Fig. 2.1 A network coding node B with H incoming links $l_1, l_2, \cdots, l_H$ and one outgoing link l_0.
(**a**) GBN-ARQ. (**b**) SR-ARQ

H information sources $X_1, X_2, \ldots, X_H$, which generate the packets to be transmitted
through the system. Note that, here what we mean the *abstract* source node is that
$X_1, X_2, \ldots, X_H$ actually represent H independent source nodes possibly distributed
at different locations. Node B is a $HISO$ coding node with packet-level network
coding. Finally, node C is a sink node receiving the coded packets from node B.
As shown in Fig. 2.1a, there is one link l_0 between nodes B and C. Between

nodes A and B, there are H links $l_1, l_2, \ldots, l_H$, each of which connects one of the sources $X_1, X_2, \cdots, X_H$ with node B. We assume that packets are transmitted over the links based on the GBN-ARQ scheme. In Fig. 2.1a, node B is a network coding node with the coding operation represented by $\star$. At node B, b_0 is the buffer for storing the packets to be transmitted to node C. By contrast, $b_1, b_2, \cdots, b_H$ are the buffers storing the packets successfully received, respectively, from the source nodes $X_1, X_2, \cdots, X_H$.

For the sake of simplicity, we assume that the transmission window of X_h is W_h packets. The transmitter of X_h sends W_h packets before it receives a feedback corresponding to the oldest packet in the transmission window. As seen in Fig. 2.1a, at the coding node B, corresponding to every link, such as link l_h, there is a buffer for storing one received packet. If each of the H links correctly receives one packet, then they are encoded, forming a packet, which is stored in b_0. We assume that b_0 can store up to W_0 packets, which is the transmission window size of node B.

Note that, above we assumed that each of the H incoming links at node B can only store at most one packet. The throughput under this assumption represents the lower-bound of the throughput achievable. If each of the H links at node B is capable of storing more packets, a higher throughput may be expected. Furthermore, if each of the H links is assumed to store infinite packets, we can obtain an upper-bound for the throughput.

In the context of the SR-ARQ, the system model considered is shown in Fig. 2.1b, which has a similar structure as that shown in Fig. 2.1a, but has the following differences: (a) Packets are transmitted over each of the links based on the SR-ARQ scheme. (b) Every link is featured with the same Round Trip Time (RTT), denoted by T. (c) For link l_h, $h = 1, 2, \ldots, H$, the receiver has a receiving window of size W_h, which is the same as the corresponding transmission window's size. When comparing Fig. 2.1b to a, we can see that the system model for the SR-ARQ is more general than that for the GBN-ARQ.

2.2.2 Main Assumptions

In addition to the above-mentioned assumptions in the system models for the GBN-ARQ and SR-ARQ, we also adopt the following assumptions in our analysis:

(1) The system is operated in a synchronous manner. This assumption is valid since, for the packet-level network coding to work, it is required that the lower layers provide proper buffering scheme to guarantee the alignment of incoming packets of the same generation.

(2) At the sources $X_1, X_2, \cdots, X_H$, there are always packets ready to send. The nth packet of the hth source X_h is denoted as $x_h(n)$.

(3) Each of the H links has two channels: the forward channel and the feedback channel. The forward channel is assumed to be a Binary Symmetric Channel (BSC). The probability of (detectable) packet errors of the hth forward channel is denoted as p_{e_h}. We assume that the undetectable packet errors can be neglected, since this probability is very small, in comparison with the

probability of detectable errors. Therefore, p_{e_h} is simply the packet error rate
(PER). Furthermore, we assume that the feedback channel is perfect without
yielding transmission error.

(4) When operated under the GBN-ARQ or SR-ARQ scheme, a transmitter is able
to continuously transmit more than one packet over the forward channel within
one RTT. Similarly, the receiver is also able to feedback the same number
of confirmations over the feedback channel within one RTT. In this case, if
we normalize the RTT by the time duration with respect to a single packet
transmission, then, the RTT of link l_h is an integer T_h, which represents the
number of packets that the transmitter of link l_h is able to send before receiving
a feedback for the first packet from the receiver.

Furthermore, we assume that half of a RTT, i.e., $T_h/2$, is required for transmitting a specific packet from one node to another by the corresponding forward
channel. Similarly, half of a RTT is required for sending a specific confirmation
signal from one node to another by the corresponding feedback channel.

(5) For the GBN-ARQ, when each of $b_1, b_2, \cdots, b_H$ obtains one packet, expressed
as $x_1(n), x_2(n), \ldots, x_H(n)$, respectively, then, once b_0 has a free unit to store,
these packets are encoded to form a coded packet $x_0(n)$, which is immediately
stored into b_0. At the same time, the buffers of $b_1, b_2, \cdots, b_H$ are released for
storing the following packets.

By contrast, for the SR-ARQ, when all the packets $x_1(n), x_2(n), \ldots, x_H(n)$ are
ready and stored in $b_1, b_2, \cdots, b_H$, then, once b_0 has a free unit to store, these
packets are encoded to form a packet $x_0(n)$, which is immediately stored into b_0.
At the same time, the memory units storing $x_1(n), x_2(n), \cdots, x_H(n)$ are released
for storing the other packets.

(6) Furthermore, when employing the GBN-ARQ, packets stored in the buffer b_0
are always stored in the memory units with the smallest possible subscripts and
following the First-In-First-Serve (FIFS) operation principles.

(7) Buffers of the sources $X_1, X_2, \ldots, X_H$ and node C can store infinite number of
packets.

2.2.3   Transmission Operations

With the aid of the above assumptions, the operations of the GBN-ARQ or SR-
ARQ-assisted network coding system can be described as follows.

First, when controlled by the GBN-ARQ, each of the transmitters invoked is able
to transmit multiple packets within one RTT. Hence, for simplicity of description,
each RTT can be divided into a number of time-chips of duration T_c. Within each
time-chip, a transmitter can send up to one packet to its receiver. Similarly, in each
time-chip, a receiver can feed back a signal to its transmitter.

Let us specifically consider the nth time-chip. We assume that, when
observing at the start of this time-chip, source $X_h, h = 1, 2, \ldots, H$, has sent
the packets $X_h(m_h), X_h(m_h + 1), \ldots, X_h(m_h + W_h - 1)$, which are stored in

the transmission window and have not been confirmed. The packets that have been sent by node B to node C but have not been confirmed are expressed as $x_0(m_0), x_0(m_0 + 1), \ldots, x_0(m_0 + W_0 - 1)$. Then, within the nth time-chip, the operations are as follows in sequence:

- The receiver for link $l_h, h = 1, 2, \ldots, H$, at node B checks whether b_h is empty and whether the received packet, which is $x_h(m_h)$, is correct. If b_h is available for storing and $x_h(m_h)$ is correctly received, then, this receiver feeds back X_h an ACK. Otherwise, the receiver feeds back X_h a NACK.
  Similarly, for link l_0, the receiver at node C checks whether the received packets, which is $x_0(m_0)$, is correct. If it is correct, node C sends node B an ACK. Otherwise, node C rejects the packet and sends node B a NACK.
- For the transmitter of link $l_h, h = 1, 2, \ldots, H$, if it detects that $x_h(m_h)$ is positively confirmed by its receiver, it then transmits a new packet $x_h(m_h + W_h)$ in the nth time-chip, and the packets in the transmission window are updated to $x_h(m_h + 1)$, $x_h(m_h + 2), \ldots, x_h(m_h + W_h)$. Otherwise, the transmitter of link l_h retransmits all the packets in the transmission window using the nth, $(n+1)$th, $\ldots, (n+W_h-1)$th time-chips. The operations of link l_0 are similar as that of l_h, except that the window size is W_0.
- Within the nth time-chip, the coding node B also checks whether encoding is possible. Specifically, if each of $b_1, b_1, \ldots, b_H$ stores one packet and b_0 has at least one space for storing the new packet, then, the contents of $b_1, b_1, \ldots, b_H$ are encoded to form a packet, which is stored in b_0. Otherwise, no new coded packet is formed.

In the context of the SR-ARQ, each RTT is also divided into time-chips of duration T_c. In each time-chip, a transmitter of a link sends one packet into the channel, if the transmitter has packets to send in its outgoing buffer. Similarly, in each time-chip, a receiver sends an ACK into the feedback channel to confirm a correctly received packet or sends a NACK to request a retransmission for a corrupted packet. Table 2.1 describes how time is allocated in each time-chip, where $TC_1, TC_2, \ldots, TC_6$ are the ending points, in which each stage is completed. The upper half of Table 2.1 corresponds to the first half of a time-chip, while the lower half of Table 2.1 corresponds to the second half a time-chip.

As shown in Table 2.1, in each time-chip, Node B completes the tasks of "Packet Reception," "Packet Processing," "Feedback Transmission (Back to the Transmitter)," "Feedback Reception," "Feedback Processing," and "Packets Transmission" sequentially. Meanwhile, while a source is idle during the first half of each time-chip, during the second half of each time-chip, it completes the tasks of "Feedback Reception," "Feedback Processing," and "Packet Transmission" sequentially. By contrast, the sink executes "Packet Reception," "Packet Processing," and "Feedback Transmission" sequentially in the first half of each time-chip and stays idle in the second half. Specially, in the session of "Packet Processing" at Node B, the coding node detects if there are any errors in the packet received in the former session. If errors are present, Node B sends the feedback of NACK back to the transmitter in the next session; otherwise, Node B sends ACK to the transmitter in

Table 2.1 Time allocation in a time-chip

t	$[0, TC_1)$	$[TC_1, TC_2)$	$[TC_2, TC_3)$
Sources	Idle	Idle	Idle
Node B	Packet Reception	Packet Processing	Feedback Transmission
Sink	Packet Reception	Packet Processing	Feedback Transmission

t	$[TC_3, TC_4)$	$[TC_4, TC_5)$	$[TC_5, TC_6)$
Sources	Feedback Reception	Feedback Processing	Packet Transmission
Node B	Feedback Reception	Feedback Processing	Packet Transmission
Sink	Idle	Idle	Idle

the next session. Similarly, in the session of "Feedback Processing," Node B makes decision based on the feedback received in the last "Feedback Reception" session: if an ACK is received, Node B marks the corresponding packet as "Confirmed" in the SR-ARQ transmitter buffer b_0 and shifts the packets stored in the transmission window to the left to transmit new packets. By contrast, if a NACK is received, the SR-ARQ transmitter marks the corresponding packet as "To be Retransmitted," so that the corresponding packet is retransmitted, when it becomes the packet with the smallest generation number among those packets that are waiting to be retransmitted.

Let us below focus on the performance analysis of the steady-state throughput of the GBN- and SR-ARQ-assisted packet-level network coding systems in Sect. 2.3 and Sect. 2.4, respectively.

2.3 Analysis of Steady-State Throughput of Network Coding Nodes Employing GBN-ARQ

In this section, we first show that a FSM can be employed to represent the network coding system shown in Fig. 2.1a. Then, the steady-state throughput of the network coding system of Fig. 2.1a is analyzed, where the steady-state throughput is defined as the packet rate received by node C. This steady-state throughput also equals the rate that the coded packets formed by node B.

2.3.1 Finite-State Machine Modeling of Network Coding Node's Operations

Let us denote the time duration of each time-chip as T_c. Each GBN-ARQ transmitter is capable of constantly sending packets at the highest rate of $1/T_c$ packets per second, and the system is observed with respect to the time interval of T_c.

As shown in Fig. 2.1a, the system consists of three nodes and $H + 1$ links. For the sake of convenience and simplicity, let us define some variables to describe the instantaneous state of the whole system.

First, let us define a delay variable $C_h = 0, 1, \ldots, T_h$, in order to describe the delays between the original packet and its retransmitted duplicate on a link that connects the source node X_h with the coding node B. It can be shown that C_h belongs to one of the three cases.

(1) $C_h = T_h$: the corresponding transmitter operates in the normal state, and the corresponding receiver keeps receiving correct packets.
(2) $C_h < T_h$: the receiver is waiting for the expected packet to arrive from the transmitter.
(3) $C_h = 0$: A corrupted packet arrives at the receiver, which is detected by the receiver. The receiver sends a NACK back to the transmitter, and switches to the lock-up state by setting $C_h = 0$, and waits for the retransmitted duplicate of the corrupted packet.

The values of the delay variables for all the incoming links are collected into a vector $C = \{C_1, C_2, \ldots, C_h, \ldots, C_H\}$.

Second, when $C_h < T_h$, then at the end of each state transition, C_h is increased by one in order to reflect the change of the delay between a corrupted packet and its correspondingly retransmitted duplicate from one time-chip to another.

Third, let $b = \{b_1, b_2, \ldots, b_h, \ldots, b_H\}$, where $b_h = 0$ indicates that the buffer b_h is occupied, while $b_h = 1$ indicates that the buffer b_h is available for receiving new packet. Let us define a scalar $b_0, b_0 = 0, 1, 2, \ldots, W_0$, which denotes the number of packets in b_0, the buffer of the outgoing link of node B.

Fourth, let us define a vector $p = \{p_1, p_2, \ldots, p_l, \ldots, p_{T_0}\}$, where p_l takes a value in $\{0, 1\}$. Specifically, $p_l = 1$ indicates that node B needs to acknowledge a packet in the next time-chip, while $p_l = 0$ indicates that it does not. In each state transition, the content in p is shifted to the right by one element, and in the mean time, if there is a packet to transmit, p_1 is set to 1, otherwise, p_1 is set to 0.

Finally, based on the variables as above-defined, we define the $S = \{C, b, b_0, p\}$ as a state of the coding node B. Let S be a set containing all the possible states and the size of S be L. Furthermore, for convenience, let us denote the i-th element in S as $S^{(i)}$.

Based on the above definitions and assumptions, we can now analyze the state transition of the network coding node B, as detailed below.

2.3.2 State Transitions

Let us denote the state transition matrix as $P = P^0 + P^+$, where the transitions represented by P^0 do not yield throughput, while those represented by P^+ generate throughput. Let us first analyze P^0.

2.3.2.1 Calculation of P^0

When there is no new packet encoded during a transition, the probability of transition from $S^{(i)}$ to $S^{(j)}$ is denoted by $P^0_{i,j}$. In this case, the operations of each of the incoming links and that of the outgoing link of the network coding node B are independent. Therefore, $P^0_{i,j}$ can be expressed as

$$P^0_{i,j} = \prod_{h=0}^{H} f_h, \tag{2.1}$$

where f_h denotes the probability contributed by link l_h.

Let $S^{(i)} = \{C^{(i)}, b^{(i)}, b_0^{(i)}, p^{(i)}\}$, where $C^{(i)} = \{C_h^{(i)}\}$, $b^{(i)} = \{b_h^{(i)}\}$, and $p^{(i)} = \{p_h^{(i)}\}$. Let us also define $Q^{(i)}$ the number of ones in $p^{(i)}$, i.e., the number of packets on the outgoing link of the coding node. Furthermore, let us define $p_L^{(i)} = \{p_1^{(i)}, p_2^{(i)}, \ldots, p_l^{(i)}, \ldots, p_{T_0-1}^{(i)}\}$, where T_0 is the RTT of link l_0. Then, f_h for $h = 1, \ldots, H$ in (2.1) can be determined as by considering the following cases.

First, if a packet is successfully received from the hth incoming link of node B, i.e., if $C_h^{(i)} = T_h, C_h^{(j)} = T_h, b_h^{(i)} = 1, b_h^{(j)} = 0$, then, we have $f_h = \bar{p}_{e_h} = 1 - p_{e_h}$.

Second, if a packet is received in error from the hth incoming link of node B, i.e., if $C_h^{(i)} = T_h, C_h^{(j)} = 0, b_h^{(i)} = b_h^{(j)} = 1$, then, we have $f_h = p_{e_h}$.

Third, if no buffer has been changed, i.e., $C_h^{(i)} < T_h, C_h^{(j)} = C_h^{(i)} + 1, b_h^{(i)} = b_h^{(j)}$; or if an incoming packet is rejected because the incoming buffer is occupied, i.e., $C_h^{(i)} = T_h, C_h^{(j)} = 0, b_h^{(i)} = b_h^{(j)} = 0$, then in these two cases, we have $f_h = 1$.

Finally, for all the other cases, we have $f_h = 0$.

By contrast, f_0 can be determined by considering the following three situations.

First, when the network coding node B is going to receive a feedback in the next time-chip, which corresponds to $p^{(i)} = [p_L^{(i)} \ 1]$, in this case, if the number of packets stored in the outgoing buffer decreases by one in the transition resulting in $b_0^{(j)} = b_0^{(i)} - 1$, then, either the outgoing buffer still has packets to transmit, giving $b_0^{(i)} > Q^{(i)}, p^{(j)} = [1 \ p_L^{(i)}]$, or the outgoing buffer has no more packets to send, corresponding to $b_0^{(i)} = Q^{(i)}, p^{(j)} = [0 \ p_L^{(i)}]$. Consequently, we have $f_0 = \bar{p}_{e_0}$. However, if the content in the outgoing buffer remains the same before and after the transition, and the transmission on the outgoing link is re-initialized, corresponding to the case where a NACK has been received, i.e., if $b_0^{(j)} = b_0^{(i)} = 1, p^{(j)} = [100\ldots0]$, then, we have $f_0 = p_{e_0}$.

Second, if the content of the outgoing buffer remains the same before and after the transition and if there is no feedback need to be responded in the transition, corresponding to $b_0^{(j)} = b_0^{(i)}, p^{(i)} = [p_L^{(i)} \ 0]$, then, either when the outgoing buffer still has packets to transmit, yielding $b_0^{(i)} > Q^{(i)}, p^{(j)} = [1 \ p_L^{(i)}]$; or when, the outgoing buffer has no more packets to transmit, yielding $b_0^{(i)} = Q^{(i)}, p^{(j)} = [0 \ p_L^{(i)}]$, we have $f_0 = 1$.

Finally, for all the other cases, $f_0 = 0$.

2.3.2.2   Calculation of P^+

When there is a new packet generated and the outgoing buffer of node B is not overflowed, which requires to satisfy $b_0^{(i)} < W_0, b_1^{(j)} = b_2^{(j)} = \ldots = b_H^{(j)} = 1$, then P^+ can be expressed as

$$P_{i,j}^+ = \prod_{h=0}^{H} f_h \qquad (2.2)$$

where f_h for $h = 1, \ldots, H$ can be determined as follows:

First, when the hth incoming link stays in the transmission mode and the corresponding incoming buffer is available to store, which corresponds to the states, $C_h^{(i)} = C_h^{(j)} = T_h, b_h^{(i)} = b_h^{(j)} = 1$, then we have $f_h = \bar{p}_{e_h}$.

Second, when a new packet is received from the hth incoming link of node B, but the corresponding incoming buffer is full, resulting in that a NACK is sent to source X_h and the states $C_h^{(i)} = T_h, C_h^{(j)} = 0, b_h^{(i)} = 0, b_h^{(j)} = 1$; or the corresponding incoming link is in the waiting mode, giving that $C_h^{(j)} = C_h^{(i)} + 1, b_h^{(i)} = 0, b_h^{(j)} = 1$, then, we have $f_h = 1$.

Finally, for all the other cases, we have $f_h = 0$.

Correspondingly, f_0 can be determined upon considering the following cases.

First, when there is a feedback, which needs to be responded in the next time-chip: ($\boldsymbol{p}^{(i)} = [\boldsymbol{p}_L^{(i)}\ 1]$), if the number of packets in the outgoing buffer remains the same during the transition ($b_0^{(j)} = b_0^{(i)}$), and, meanwhile, if there are packets in the outgoing buffer, which is required to be transmitted ($W_0^{(i)} > Q^{(i)}, b_0^{(i)} + 1 > Q^{(i)}, \boldsymbol{p}^{(j)} = [1\ \boldsymbol{p}_L^{(i)}]$), and the outgoing link is full ($W_0^{(i)} = b_0^{(i)} + 1 = Q^{(i)}, \boldsymbol{p}^{(j)} = [0\ \boldsymbol{p}_L^{(i)}]$), then, we have $f_0 = \bar{p}_{e_0}$; or if the transmission on the outgoing link is re-initialized, yielding $b_0^{(j)} = b_0^{(i)}, \boldsymbol{p}^{(j)} = [100\ldots0]$, then we have $f_0 = p_{e_0}$.

Second, when there is no feedback required to be responded in the next time-chip ($b_0^{(j)} = b_0^{(i)} + 1, \boldsymbol{p}^{(i)} = [\boldsymbol{p}_L^{(i)}\ 0]$), and, meanwhile, either if there are packets in the outgoing buffer that need to be transmitted ($W_0^{(i)} > Q^{(i)}, b_0^{(i)} + 1 > Q^{(i)}, \boldsymbol{p}^{(j)} = [1\ \boldsymbol{p}_L^{(i)}]$), or if the outgoing buffer is full ($W_0^{(i)} = b_0^{(i)} + 1 = Q^{(i)}, \boldsymbol{p}^{(j)} = [0\ \boldsymbol{p}_L^{(i)}]$), then, we have $f_0 = 1$.

Finally, for all the other cases, we have $f_0 = 0$.

When both $P_{i,j}^+$ and $P_{i,j}^0$ for all $i, j \in \mathcal{L} = \{1, 2, \ldots, L\}$ prepared, the probability transition matrix $\boldsymbol{P} = (P_{i,j})$ can be formed as $\boldsymbol{P} = \left(P_{i,j} = P_{i,j}^+ + P_{i,j}^0\right)$. The probability transition matrix $\boldsymbol{P}$ can also be decomposed into the form of $\boldsymbol{P} = \boldsymbol{P}^+ + \boldsymbol{P}^0$, where $\boldsymbol{P}^+ = \left(P_{i,j}^+\right)$ and $\boldsymbol{P}^0 = \left(P_{i,j}^0\right)$. Below we analyze the steady-state throughput of the network coding node employing GBN-ARQ.

2.3.3   Steady-State Throughput

Let $P_i(m)$ denote the probability that the state of node B is $S^{(i)}$ at the beginning of the mth time-chip. Let $p(m) = [P_1(m), P_2(m), \cdots, P_L(m)]^T$, where L is the size of the set S. Explicitly, we have

$$\sum_{l=1}^{L} P_l(m) = 1. \tag{2.3}$$

Then, using the *law of total probability*, we can express the probability $P_j(m+1)$ as

$$P_j(m+1) = \sum_{i=1}^{L} P_{i,j} P_i(m), \ 0 \le j \in S \tag{2.4}$$

where $P_{i,j} = P_{i,j}(m)$ denotes the transition probability from state $S^{(i)}$ at $t = mT_c$ to state $S^{(j)}$ at $t = (m+1)T_c$, which has been analyzed in Sect. 2.3.2. Note that, the transition probabilities are time-invariant and independent of m. Alternatively, (2.4) with all $j \in S$ can be expressed in vector form as

$$p(m+1) = P^T p(m) \tag{2.5}$$

which is a recursive equation. Hence, given $p(0)$, $p(m)$ can be expressed as

$$p(m) = \left(P^T\right)^m p(0), \ m = 1, 2, \ldots \tag{2.6}$$

As shown in (2.3), the sum of each row of P equals one. Hence, P^T is a left stochastic matrix [22], whose limit of $\lim_{m \to \infty} (P^T)^m$ exists, according to the *Perron-Frobenius theorem* [22, 23]. Therefore, when $m \to \infty$, the Markov process becomes stationary [24] and yields

$$p(m+1) = p(m) = \pi \tag{2.7}$$

where $\pi = [\pi_1, \pi_2, \cdots, \pi_L]^T = \lim_{m \to \infty} p(m)$ and $\sum_{1 \le i \le L} \pi_i(m) = 1$. Therefore, letting $m \to \infty$ and applying (2.7) into (2.5), we have

$$\pi = P^T \pi \tag{2.8}$$

under the constraint of $\sum_{1 \le i \le L} \pi_i(m) = 1$. Equation (2.8) shows that π is a right eigenvector of matrix P^T corresponding to the eigenvalue one. Therefore, the solution to π can be obtained with the aid of the methods for solving the eigenvector problem [22, 23].

Finally, when reaching the steady-state, the throughput of the *H*ISO network coding system using GBN-ARQ can be measured by the rate that packets in buffers $b_1, b_2, \cdots, b_H$ are successfully encoded and forwarded to buffer b_0. When this rate is normalized by the length of a time-chip, which we denote as T_c, the throughput of the network coding system equals the probability that new coded packets are formed and forwarded to b_0 in each time-chip. According to the operation principles as detailed in Sect. 2.3.2, throughput is generated, only when the state transitions of node *B* result in the non-zero transition probabilities of $\left\{P_{i,j}^+\right\}$, which are the non-zero elements in the transition matrix P^+. Therefore, the steady-state throughput of the *H*ISO network coding system using GBN-ARQ can be expressed as

$$R_c = \sum_{i=1}^{L} \pi_i \sum_{j=1}^{L} P_{i,j}^+ \tag{2.9}$$

where the throughput R_c is given in the unit of the number of packets per time-chip. When each of the link in the system has the same RTT of T, where T is the number of time-chips in one RTT, the throughput can also be express as $R_{RTT} = R_c T$, where R_{RTT} is the steady-state throughput given in the unit of the number of packets per RTT.

2.4   Analysis of Steady-State Throughput of Network Coding Nodes Employing SR-ARQ

In this section, we first show how a FSM can be employed to represent the network coding system employing the SR-ARQ, as shown in Fig. 2.1b. Then, the steady-state throughput of the network coding system employing SR-ARQ is analyzed. As done for the GBN-ARQ, here, the steady-state throughput is defined as the packet rate received by node *C*, which, in the steady-state, equals the rate that the coded packets formed by node *B*.

2.4.1   Finite-State Machine Modeling of Network Coding Node's Operations

As for the GBN-ARQ, let us denote the time duration of each time-chip as T_c. When operated under the SR-ARQ, a transmitter is capable of continuously sending packets at the highest rate of $1/T_c$, and the system is observed with the time interval of T_c. As shown in Fig. 2.1b, the system consists of three nodes and $H+1$ links. For the sake of convenience and simplicity, below we define some variables, in order to describe the instantaneous state of the whole system. Specially, in order to keep track of the packet transmissions on each link, we introduce the variables referred to as the *counters*, which are explained in detail below.

2.4.1.1   Packet Transmission Modeling with Counters

Each counter is used to describe the state of a packet transmitted from a certain generation on a link. A counter indicates the remaining number of time-chips needed for a packet to complete the transmission on a link. On each link, multiple counters are needed, if packets from multiple generations are transmitted simultaneously. Below let us explain the concept of counters in detail.

First, let W_0 be the counters needed in order to track the transmissions on the outgoing link l_0, as the outgoing buffer of the SR-ARQ transmitter has a number of W_0 units, meaning that there are at most W_0 packets from different generations being transmitted on the outgoing link l_0 at the same time. Let the counters corresponding to link l_0 be denoted by the set $C_0 = [C_0^1 C_0^2 \ldots C_0^{m_0} \ldots C_0^{W_0}]$. The count-down counter $C_0^{m_0}$ keeps tracking the time elapsing from the moment when the packet in the m_0th unit of the outgoing buffer leaves from the transmitter (node B) to the moment when the corresponding feedback is received by node B.

By contrast, for the hth incoming link, $2W_h$ number of counters are required, in order to track the transmission state on the hth incoming link. The reason behind can be explained as follows. First, even when all the W_h packets in the incoming buffer b_h have been confirmed, b_h may still be fully occupied, if the buffer of the SR-ARQ transmitter of link l_0 is full. Meanwhile, there might be a total W_h number of packets from different generations being transmitted or waiting for the feedbacks on the link l_h. Therefore, in order to track the transmission states of all the packets that have been confirmed and that being in transient on link l_h, a total $2W_h$ number of counters are necessary for link l_h. In other words, in the worst case, the longest causal delay of the incoming link l_h has a length of $2W_h$ time-chips.

Let the counter set corresponding to link $l_h, 1 \le h \le H$, be denoted by $C_h = [C_h^1 C_h^2 \ldots C_h^{m_h} \ldots C_h^{2W_h}]$. The count-down counter $C_h^{m_h}, 1 \le h \le H$, keeps tracking of the time elapsing from the **Event 1** to the **Event 2**, which are defined as:

Event 1: The feedback for the m_hth packet in the incoming buffer of link l_h is transmitted by node B.
Event 2: The packet requested in **Event 1** arrives at the SR-ARQ receiver of node B.

It can be shown that $C_h^{m_h}, 0 \le h \le H$, takes value from the set $\{-1, 0, 1, \ldots, T_h\}$ and the values of $C_h^{m_h}$ have the following meanings:

- When $C_h^{m_h} = -1$, it indicates that the corresponding counter is inactive. In this case, for a receiver, it means that the corresponding packet has not been requested; while for a transmitter, it means that the corresponding packet has not been sent.

- When $C_h^{m_h} = 0$, it indicates that the corresponding packet has been "Confirmed." In this case, for a receiver, it means that the corresponding packet has been received correctly; while for a transmitter, it means that the corresponding packet has been acknowledged by an ACK.
- $0 < C_h^{m_h} \leq T_h$, for a transmitter, it indicates that the transmitter is waiting for the corresponding packet to be acknowledged either by an ACK or by a NACK. By contrast, for a receiver, it indicates that the receiver is waiting for the requested packet to arrive.

As an example for the above, for the transmitter connected to link l_0, $C_0^{m_0} = T_0$ indicates that the corresponding packet has just been sent by the transmitter into the forward channel, while $C_0^{m_0} = 1$ indicates that the feedback is going to arrive at the transmitter in the next time-chip. By contrast, for a receiver connected to the incoming links $l_h, 1 \leq h \leq H$, $C_h^{m_h} = T_h$, indicates that a feedback has just been sent by the transmitter into the feedback channel to request the corresponding packet, while $C_h^{m_h} = 1$ indicates that the expected packet is going to arrive at the transmitter in the next time-chip.

The value of the counter corresponding to a given generation is changed accordingly following the following rules. When a corrupted packet arrives at the receiver, the receiver detects the errors in the packet and sends a NACK back to the transmitter. At the same time, the corresponding counter is set to $C_h^{m_h} = T_h$, and the receiver waits for the retransmitted duplicate of the corrupted packet. Furthermore, as we mentioned previously, the time duration between an error is detected and the arrival of the retransmitted duplicate is T_h, namely a RTT. Therefore, at the end of each time-chip with $0 < C_h^{m_h} \leq T_h$, $C_h^{m_h}$ decreases by 1. In this way, the delay of T_h between a corrupted packet and its corresponding retransmitted duplicate can be simulated.

There are some facts, which should be noted. First, for a normal SR-ARQ without the network coding function, W_h counters are sufficient for tracking the state of the transmitter and receiver, since once a packet is successfully received, this packet is automatically released from the receiver. For example, in order to track the state transitions on link l_0, only W_0 counters are needed, as the SR-ARQ receiver at sink node C has infinite buffer capacity and therefore is always ready to release the right-most packets received in the receiver's buffer. By contrast, at the SR-ARQ-assisted network coding node, even a packet is correctly received, it is not guaranteed that this packet can be passed to the network coding encoder immediately, since the buffer b_0 of the SR-ARQ transmitter of link l_0 might be full. Therefore, a number of $2W_h$ counters are required for links $l_h, 1 \leq h \leq H$. Second, normally, a network starts from an idle state, which means that there are no packets being transmitted on any of the links at the beginning. The SR-ARQ model considered in this chapter does not cover this special case, since we focus on the steady-state analysis. Furthermore, at node B, the counting procedure for a counter at the receiver is different from that of the transmitter in the following aspects:

(a) **Number of Counters**—For a SR-ARQ receiver connected to an incoming link l_h, $2W_h$ counters are needed, while for the SR-ARQ transmitter connected to an outgoing link, only W_0 counters are needed.

(b) **Commerce of Down-Counting**—At a SR-ARQ receiver, counters for an incoming link starts counting down, when a feedback leaves node B, while counters for an outgoing link starts counting down, when a packet leaves node B.

2.4.2 State Transitions

In the process of packet transmission modeled with the aid of counters, it can be seen that the instantaneous state of the system can be described by the states of the counters, as well as the number of packets in the outgoing buffer. Therefore, in order to model the activities of the system, let us denote the number of packets in the outgoing buffer b_0 as k. With all the $H + 1$ sets of counters $C_0, C_1, \ldots, C_h, \ldots, C_H$ and k, all the system parameters can be determined, including the status of the incoming and the outgoing buffers, as well as the transmissions in the forward and feedback channels.

All the possible states of the system form a state set, which we denote as $\mathcal{S}$. If we denote the ith element in this set as $S^{(i)} = [C_0^{(i)}, C_1^{(i)}, \ldots, C_h^{(i)}, \ldots, C_H^{(i)}, k^{(i)}]$, where $C_0^{(i)}, C_1^{(i)}, \ldots, C_h^{(i)}, \ldots, C_H^{(i)}, k^{(i)}$ are the counters and the number of packets in the outgoing buffer in this state, respectively. On the other hand, let us denote the state of the system at the beginning of the mth time-chip as $S(m)$. Also, let us denote the state of the system at the end of the first half of the mth time-chip as $\tilde{S}(m)$. Both $S(m) \triangleq [C_0(m), C_1(m), \ldots, C_h(m), \ldots, C_H(m), k(m)]$ and $\tilde{S}(m) \triangleq [C_0'(m), C_1'(m), \ldots, C_h'(m), \ldots, C_H'(m), k'(m)]$ take their values from the set $\mathcal{S}$. In the rest of this section, we aim to determine the probability $P_{i,j}(m)$, which is the probability of the transition from the state $S(m) = S^{(i)}$ to the state $S(m + 1) = S^{(j)}$. As we will see, the probability $P_{i,j}(m)$ is time-invariant. Therefore, we drop the time-chip indices for convenience and denote it as $P_{i,j}$. Similarly, we have $S \triangleq [C_0, C_1, \ldots, C_h, \ldots, C_H, k]$ and $\tilde{S} \triangleq [C_0', C_1', \ldots, C_h', \ldots, C_H', k']$. Specially, if at the end of the first half of a transition, the hth incoming link of node B is in the state $S^{(i)}$, then the corresponding counters are denoted as $C_h'^{(i)}$.

We will first calculate the probability $P_{i,j}$ under the condition that there are no new packets generated in the transition. Then, we calculate the probability under the condition that there are packets drawn from the incoming buffers to form new outgoing packets. Let us collect the probabilities of the former kind into the matrix denoted by $P^0 = \{P_{i,j}^0\}$ and collect the probabilities of the latter kind into the matrix denoted by $P^+ = \{P_{i,j}^+\}$. Furthermore, let the number of new packets generated during each transition be denoted by $\gamma = \{\gamma_{i,j}\}$, where $\gamma_{i,j}$ is the number of new packets generated during the transition from $S^{(i)}$ to $S^{(j)}$ under the condition that there are new packets formed in the transition. Below we calculate P^0, P^+, and γ.

2.4.2.1   Necessary Conditions for a State

First of all, let us determine which states are included in the state set $\mathcal{S}$, in other words, which states are legitimate. For a transition from $S^{(i)}$ to $S^{(j)}$ to be legitimate, it is required that both of the states $S^{(i)}$ and $S^{(j)}$ are legitimate. Otherwise, both the probabilities of $P^0_{i,j}$ and $P^+_{i,j}$ are equal to zero, and, therefore, there is no need to calculate. Consequently, those combinations of $[C^{(i)}_0, C^{(i)}_1, \ldots, C^{(i)}_h, \ldots, C^{(i)}_H, k^{(i)}]$ that do not fulfill the necessary conditions can be omitted from the state set $\mathcal{S}$. In detail, for a state to be legitimate, it is required that the following conditions are fulfilled.

(1) On each of the incoming links, each packet has a unique generation number. This means that, in any C_h, there exists no more than one counter having equal value, except those counters with a value of zero or one.
(2) There are at most W_h packets being transmitted on an incoming link l_h, meaning that for any C_h, there are at most W_h counters, which take the values other than 0 and -1.
(3) An incoming buffer follows the FIFS principle and records the active packets on the right-most unit. In other words, if there is an -1 element in C_h, it must be the left-most element.
(4) At the SR-ARQ transmitter of node B, once a packet is confirmed that it has been delivered successfully, there is no need to track it any more. Correspondingly, there does not exist a single zero element or consecutive zero elements at the right-most part of C_0.
(5) In the outgoing buffer of node B, the total number of packets is always equal to or larger than the number of active packets, yielding that k is always larger than or equal to the number elements in C_0 that are not -1.
(6) For the SR-ARQ transmitter at node B, whenever it is possible, the transmitter always tries to transmit or retransmit the oldest packet that needs to be transmitted. This packet can either be a new packet or a packet that has been requested for a retransmission. Under this assumption, if there is no $C^{m_0(i)}_0 = 1$, $1 \leq m_0 \leq W_0$, and if k is larger than the number of elements in C_0 that are other than -1, then, in an outgoing buffer, for the right-most $C^{m_0(i)}_0 = -1$, the corresponding $C^{m_0(j)}_0 = T_0$.

2.4.2.2   The Calculation of P^0

When there is no new packet encoded during a transition, the probability from $S^{(i)}$ to $S^{(j)}$ is denoted by $P^0_{i,j}$. There are two possible cases for a zero-throughput transition. The first case is that the outgoing buffer is fully occupied in both the state $S^{(i)}$ and the state $S^{(j)}$, and in the meantime, there is no packet cleared from the outgoing buffer during this transition. The second case is that there are available buffer units in the outgoing buffer but there is at least one right-most packet in an incoming link remains unconfirmed during this transition. Therefore, the following necessary conditions are satisfied, when no throughput is generated during the transition. First, for every $C_h^{m_h(i)} > 1$ in $S^{(i)}$, the corresponding counter $C_h^{m_h(j)}$ in $S^{(j)}$ must equal $C_h^{m_h(i)} - 1$. Second, counters corresponding to the packets that have been positively confirmed stay unchanged. In other words, for every $C_h^{m_h(i)} = 0$ in $S^{(i)}$, the corresponding counter $C_h^{m_h(j)}$ in $S^{(j)}$ must stay 0. Furthermore, it can be understood that, when no throughput is generated during a transition, there is no interaction between the stage before the network encoder and the stage after it. Therefore, $P_{i,j}^0$ is the product of the probabilities of the state transitions occurred before the network encoder and that occurred after the encoder. Let $P_{i,j}^0$ be expressed by

$$P_{i,j}^0 = \prod_{h=0}^{H} f_h \tag{2.10}$$

where the value of f_h for $h = 1, \dots, H$ is determined by the conditions listed below, which are broken into two cases, and explained as follows:

Case 1   The outgoing buffer is fully occupied at both state $S^{(i)}$ and state $S^{(j)}$, and in the meantime, there are no packets cleared from the outgoing buffer during this transition. In order to qualify for this category, the system must fulfill the following conditions:

(1) The outgoing buffer is full. Then, the number of packets in the outgoing buffer equals the length of the outgoing buffer, yielding $k = W_0$.
(2) There is no inactive packets in the outgoing buffer, i.e., every packet in the outgoing buffer is being transmitted or waiting to be confirmed, corresponding to $\forall C_0^{m_0}, C_0^{m_0} \geq 0$ ($1 \leq m_0 \leq W_0$).

Based on the above observation, let us first determine f_h for $1 \leq h \leq H$. First, if an incoming packet is received in error, then a NACK is sent back to the transmitter. In this case, for an incoming buffer, if $C_h^{m_h(i)} = 1$ and correspondingly $C_h^{m_h(j)} = T_h$, then we have $f_h = p_{e_h}$.
Second, if an incoming packet in the buffer b_h is correctly received, then, the corresponding counter is set to 0. In this case, for an incoming buffer, if $C_h^{m_h(i)} = 1$ and correspondingly $C_h^{m_h(j)} = 0$, then we have $f_h = \bar{p}_{e_h} = 1 - p_{e_h}$.
Third, if there is a chance to transmit a new packet or retransmit a packet that has been requested, the transmitter transmits a packet. In this case, for an incoming buffer, if there does not exist a $C_h^{m_h(i)} = 1$ and there are a number of buffer units in $C_h^{'(i)}$, which have the values other than -1, for the $C_h^{m_h(i)} = -1$ with the largest h, $C_h^{m_h(j)} = T_h$, then we have $f_h = 1$.
Finally, in any other cases, if the necessary conditions are fulfilled, then we have $f_h = 1$. Otherwise, $f_h = 0$.
By contrast, the probability f_0 can be determined by considering the following cases. First, if the SR-ARQ transmitter of node B receives a NACK, it retransmits

the corresponding packet. In this case, considering the outgoing buffer, if there
is a $C_0^{mo(i)} = 1, 1 \leq w_0 \leq W_0$, and the corresponding $C_0^{mo(j)} = T_0$, we then have
$f_0 = p_{e_0}$.
Second, if the SR-ARQ transmitter of node B receives an ACK, it sets the
corresponding counter to 0, indicating that the packet has been confirmed. In such
a case, then considering an outgoing buffer, if there is a $C_0^{mo(i)} = 1, 1 \leq w_0 < W_0$,
and the corresponding $C_0^{mo(j)} = 0$, we have $f_0 = \overline{p}_{e_0}$.
Finally, in any other cases, if the necessary conditions of a state are fulfilled, we
have $f_0 = 1$. Otherwise, we have $f_0 = 0$.

Case 2   Outgoing buffer of node B has spaces for storing new packets, but, in the
buffers of the incoming links of node B, there is at least one right-most packet
remaining unconfirmed during the transition. In order to qualify the probability
for this category, the system must fulfill the following conditions:

(1) There are available buffer units in the outgoing buffer, meaning that $k < W_0$.
(2) No right-most packets in any incoming buffers are confirmed during the
    transition. Alternatively, although some of the right-most packets in the
    incoming buffers are confirmed during the transition, at least one of them is
    received in error. In this case, there does not exist any $C_h^{W_h(i)} = 1$ $(1 \leq h \leq H)$,
    or for at least one of the $C_h^{W_h(i)} = 1$ $(1 \leq h \leq H)$, the corresponding
    $C_h^{W_h(j)} = T_h$.

Based on these conditions, we can now first determine $f_h, 1 \leq h \leq H$. First, if
an incoming packet is received in error, the corresponding counter is set to T_h. In
this case, for an incoming buffer, if $C_h^{m_h(i)} = 1$ and correspondingly $C_h^{m_h(j)} = T_h$,
we have $f_h = p_{e_h}$.
Second, if an incoming packet in the buffer $\boldsymbol{b}_h$ is correctly received, then the
corresponding counter is set to 0. In this case, when considering an incoming
buffer, if $C_h^{m_h(i)} = 1$ and the corresponding $C_h^{m_h(j)} = 0$, we have $f_h = \overline{p}_{e_h}$.
Third, if there is a chance to transmit a new packet or retransmit a packet that has
been requested, the transmitter transmits a packet. In this case, for the buffer $\boldsymbol{b}_h$,
if there does not exist a $C_h^{m_h(i)} = 1$ and there are a number (which is less than W_h)
of buffer units in $\boldsymbol{C}_h^{'(i)}$ that have the values other than -1, for the $C_h^{m_h(i)} = -1$
with the largest h, $C_h^{m_h(j)} = T_h$, we have $f_h = 1$.
Finally, in any other cases, if the necessary conditions are fulfilled, we have $f_h = 1$. Otherwise, if the necessary conditions are not satisfied, we have $f_h = 0$.
Similarly, the probability f_0 can be determined as follows. First, if the SR-ARQ
transmitter of the outgoing link of node B receives a NACK, it retransmits the
corresponding packet. Correspondingly, when $k^{(j)} = k^{(i)}$, if there is a $C_0^{mo(i)} = 1, 1 < w_0 \leq W_0$, and, correspondingly, $C_0^{mo(j)} = T_0$, then we have $f_0 = p_{e_0}$.
Second, if the SR-ARQ transmitter of the outgoing link of node B receives an
ACK that corresponds to the packet that is not the right-most packet in the
outgoing buffer, the SR-ARQ transmitter marks the corresponding packet as

confirmed. In this case, in b_0, when $k^{(j)} = k^{(i)}$, if there is a $C_0^{m_0(i)} = 1, 1 < w_0 < W_0$, and the corresponding $C_0^{m_0(j)} = 0$, then we have $f_0 = \bar{P}_{e_0}$.

By contrast, if the SR-ARQ transmitter of the outgoing link of node B receives an ACK that corresponds to the packet, which is the right-most packet in the outgoing buffer, the SR-ARQ transmitter releases all the packets in the buffer that have been confirmed, and, simultaneously, shifts the rest packets to the right, and reduces all the rest of the counters by 1. In this case, when considering b_0, for $k^{(j)} < k^{(i)}$, if $C_0^{m_0(i)} > 1, m_0(i) > W_0 - k^{(j)}$, let us denote the number of zeros and the zero on the right-most part of $C_h^{(i)}$ as $g_h^{(i)}$. Then, if $C_0^{m_0(j)} = C_0^{m_0 - g_h^{(i)}(i)} - 1$, we have $f_0 = 1$.

Third, if the SR-ARQ transmitter of the outgoing link of node B receives an ACK that corresponds to the packet, which is the right-most packet in the outgoing buffer, the SR-ARQ transmitter then releases all the packets in the buffer that have been confirmed, shifts all the rest of packets to the right, and marks the packets on the left of the buffer as inactive packets. In this case, for b_0, when $k^{(j)} < k^{(i)}$, if $C_0^{m_0(i)} > 1, m_0(i) \leq W_0 - k^{(j)}$, let us denote the number of zeros and the zero on the right-most part of $C_h^{(i)}$ as $g_h^{(i)}$. If $C_0^{m_0(j)} = C_0^{m_0 - g_h^{(i)}(i)} - 1$ and $C_0^{m_0 - g_h^{(i)}(i)} = -1$, we then have $f_0 = 1$.

Fourth, if the SR-ARQ transmitter of the outgoing link of node B receives an ACK that corresponds to the packet, which is the right-most packet in the outgoing buffer, the SR-ARQ transmitter then releases all the packets in the buffer that have been confirmed, and shifts all the rest packets to the right. In this case, when considering the outgoing buffer and for $k^{(j)} < k^{(i)}$, if $C_0^{W_0(i)} = 1$, let us denote the number of zeros and the zero at the right-most part of $C_h^{(i)}$ as $g_h^{(i)}$. If one is in the right-most position and all the right-most zeros disappear in C_0^j, and if $C_0^{m_0(j)} = C_0^{m_0 - g_h^{(i)}(i)} - 1$ as well as $C_0^{m_0 - g_h^{(i)}(i)} = -1$, then we have $f_0 = \bar{P}_{e_0}$.

Finally, for all the other cases, if the necessary conditions are fulfilled, then $f_0 = 1$, otherwise, $f_0 = 0$.

2.4.2.3   Calculation of P^+

When there is a new packet encoded during a transition from $S^{(i)}$ to $S^{(j)}$, we denote the probability of such case as $P_{i,j}^+$. New packets can be generated either in the first half of a time-chip or in the second half of it, or in both of the halves. In particular, when there are available buffer units in the outgoing buffer, if the transmissions complete for all the incoming packets of a generation in the first half of a time-chip, new packets are generated immediately after the transmissions complete. Otherwise, when the outgoing buffer is fully occupied in the first half of a time-chip, and when there are completed generations that have all the incoming packets ready in the incoming buffers, then, if there are packets cleared in the second half of the time-chip, new packets are generated. Therefore, the following two conditions must be fulfilled for generating throughput:

(1) In the current state, if the outgoing buffer is full, the right-most packets in all the incoming buffers must be either correctly received or have the potential to be correctly received in the next time-chip. In this case, if $k^{(i)} = W_0$, then $C_0^{m_0(i)}$ must equal 1 and the right-most counter for each $C_h^{m_h(i)}$, $1 \leq h \leq H$, must equal 1 or 0.
(2) In the state $S^{(i)}$, node B has the potential to have new packets correctly received from the incoming links in the next time-chip. In this case, if $k^{(i)} < W_0$, the right-most non-zero counter for each $C_h^{m_h(i)}$, $1 \leq h \leq H$, must equal 1.

For the sake of convenience, as mentioned before, let us divide each time-chip into two parts and denote the state of the system at the end of the first half of the mth time-chip before network coding as $\tilde{S}(m)$, which has the following properties:

(1) $\tilde{S}(m)$ takes its value from the set $\mathcal{S}$.
(2) The state $\tilde{S}(m)$ needs to fulfill all the requirements for a state of this category, except that, when the outgoing buffer is not full, there can be consecutive zeros in the right-most position in $C_h^{m_h}$ for $1 \leq h \leq H$.

Let the $\tilde{S}(m)$ having the lth value in the set $\mathcal{S}$ be denoted by $\tilde{S}^{(l)}(m)$. Let $P_{i,v}^{+'}(m)$ denote the probability for the transition from the state at the beginning of the mth time-chip, $S(m) = S^{(i)}$, to the state at the end of the first half of the mth time-chip $\tilde{S}(m) = \tilde{S}^{(v)}$, which is the vth state in the state set $\mathcal{S}$. As we will see, the probability $P_{i,l}^{+'}(m)$ is time-invariant. Therefore, we can drop the time indices and denote the corresponding variables as $P_{i,l}^{+'}$, $\tilde{S}$, and $\tilde{S}^{(v)}$. Let $\gamma_{l,j}$ denote the number of packets generated in the transition from the state $\tilde{S}^{(v)}$ to the state $S^{(j)}$. Let us also denote the number of consecutive zeros on the right-most side of the counter set $C_h^{(i)}$ as $g_h^{(i)}$. Then, we have the following properties:

(1) When the coding node receives an ACK from its feedback channel, then, $\gamma_{l,j} = \min(\{g_h^{(i)}\}, M_0 - k^{(i)})$, where $1 \leq h \leq H$.
(2) When the coding node receives a NACK from its feedback channel, then, $\gamma_{l,j} = \min(\{g_h^{(i)}\}, M_0 - k^{(i)} + g_0^{(i)})$, where $1 \leq h \leq H$.

With all the variables as above-defined, $P_{i,l}^{+'}$ can be expressed as

$$P_{i,l}^{+'} = f_0 \prod_{\forall h \neq 0, m_h} f'_{h,m_h} \tag{2.11}$$

where the values of f'_{h,m_h}, $h = 1, \ldots, H$, and f_0 can be determined as follows.

Specifically, in the context of f'_{h,m_h} for $h = 1, \ldots, H$, we can have the following results. First, for a counter that is in the waiting mode, its value decreases by 1 in each time-chip. In this case, if $1 \leq C_h^{m_h(i)} \leq T_h$, $C_h^{m_h(v)} = C_h^{m_h(i)} - 1$, then $f'_{h,m_h} = 1$.

Second, if an incoming packet has been received correctly, the incoming link starts retransmitting a packet that has been requested before or transmitting a new packet. In this case, if $C_h^{m_h(i)} = 1, C_h^{m_h(v)} = 0$ and the right-most counter equaling -1 inside the transmission window changes to T_h in the state $S^{(v)}$, then $f'_{h,m_h} = \bar{p}_{e_h}$.

Third, if a packet has been received in error, the SR-ARQ receiver requests a retransmission. In this case, if $C_h^{m_h(i)} = 1, C_h^{m_h(v)} = T_h$, then $f'_{h,m_h} = p_{e_h}$.

Finally, for all the others cases, we have $f'_{h,m_h} = 0$.

In the context of f_0, because the first half-transition only considers the change in the left half of the coding node, the counter C_0 must stay the same during the second half of the transition. Then, if $C_0^{(v)} = C_0^{(i)}$, then we have $f_0 = 1$. Otherwise, $f_0 = 0$.

Let us now consider the probability, $P_{l,i}^{+''}$, of the transition starting from $S^{(v)}$ at the end of the first half to the state $S^{(j)}$ at the beginning of the next time-chip. It can be shown that $P_{i,l}^{+''}$ can be expressed as

$$P_{l,i}^{+''} = \prod_{\forall h,m_h} f''_{h,m_h} \tag{2.12}$$

where the values of $f''_{h,m_h}, h = 1, \ldots, H$, can be determined by considering the following two cases.

First, in the cases that the contents of each incoming buffer are shifted to the right by the number of new packets generated, for every $C_h^{m_h(v)}, m_h \leq W_h - \gamma_{l,j}$, and $C_h^{m_h+\gamma_{l,j}(j)} = C_h^{m_h(v)}$, we have $f''_{h,m_h} = 1$. Note that, the count-down of a counter in the waiting mode is carried out during the first half of each time-chip.

Second, in the cases that the packets that have been confirmed at the end of the first half of the time-chip are released, for $C_h^{m_h(j)} = 0, W_h - \gamma_{l,j} \leq m_h \leq W_h$, we have $f''_{h,m_h} = 1$.

Finally, in all the others cases, we have $f''_{h,m_h} = 0$.

By contrast, f''_0 in (2.12) can be determined as follows.

(1) If $C_0^{W_0(v)} = 1$, and the following conditions are fulfilled, we have $f''_0 = \bar{p}_{e_0}$.

   (a) Confirmed packets are released and the rest of the packets in the buffer are shifted to the right: for $1 < C_0^{m_0(v)} \leq T_0, C_0^{m_0(j)+\gamma_{l,j}-g_h^{(i)}} = C_0^{m_0(v)} - 1$.
   (b) A new packet or the earliest requested erroneous packet is transmitted: for the right-most $C_0^{m_0(v)} = -1$ in $\tilde{S}^{(v)}, C_0^{m_0(j)+\gamma_{l,j}-g_h^{(i)}} = T_0$ in $S^{(j)}$.
   (c) The packets that are already confirmed but not at the right-most side of the buffer—therefore are not released—are shifted to the right: for $C_0^{m_0(v)} = 0$, $C_0^{m_0(j)+\gamma_{l,j}-g_h^{(i)}} = 0$.
   (d) Update the number of packets in the outgoing buffer: $k^{(j)} = k^{(i)} + \gamma_{l,j} - g_h^{(i)}$.
   (e) Shift the packets in the outgoing buffer and set the empty buffer units to inactive: the left-most $W_0 - k^i - \gamma_{l,j} + g_h^{(i)}$ counters equal -1.

(2) If $C_0^{W_0(v)} = 1$, and the following conditions are fulfilled, then $f_0'' = p_{e_0}$.

  (a) Confirmed packets are released and the rest of the packets in the buffer are shifted to the right: for $1 < C_0^{m_0(v)} \leq T_0$, $C_0^{m_0(j)} = C_0^{m_0(v)} - 1$.
  (b) A retransmission is sent for the just received NACK: for $C_0^{W_0(v)} = 1$ in $\tilde{S}^{(v)}$, $C_0^{W_0(j)} = T_0$ in $S^{(j)}$.
  (c) The packets that are already confirmed but not at the right-most side of the buffer—therefore are not released—are shifted to the right: for $C_0^{m_0(v)} = 0$, $C_0^{m_0(j)+\gamma_{l,j}} = 0$.
  (d) Update the number of packets in the outgoing buffer: $k^{(j)} = k^{(i)} + \gamma_{l,j}$.
  (e) Shift the packets in the outgoing buffer and set the empty buffer units to inactive: the left-most $M_0 - k^i - \gamma_{l,j}$ counters equal -1

(3) If $C_0^{W_0(v)} > 1$, and the following conditions are fulfilled.

  (a) for a counter that is in the waiting mode, its value decreases by 1 in each time-chip: $\forall C_0^{m_0(v)}$ with $1 < C_0^{m_0(v)} \leq T_0$, we have $C_0^{m_0(j)} = C_0^{m_0(v)} - 1$;
  (b) the counters corresponding to the packets that have been 'Confirmed' stay unchanged: for $C_0^{m_0(v)} = 0$, $C_0^{m_0(j)} = 0$;
  (c) update the number of packets in the outgoing buffer: $k^{(j)} = k^i + \gamma_{l,j}$;
  (d) if there are unprocessed retransmission requests, the SR-ARQ transmitter retransmits the oldest erroneous packet in the transmission windows; if there is not any unprocessed retransmission request, the SR-ARQ transmitter transmits a new packet: for the right-most $C_0^{m_0(v)} = -1$, $m_0 \leq k^{(j)}$ in $\tilde{S}^{(v)}$, $C_0^{m_0(j)} = T_0$ in $S^{(j)}$;
  (e) shifting the packets in the outgoing buffer and setting the empty buffer units to inactive: the left-most $W_0 - k^i - \gamma_{l,j}$ counters equal -1;

  and then

  (a) if the SR-ARQ transmitter receives an ACK, i.e., given $C_0^{m_0(v)} = 1$ in $\tilde{S}^{(v)}$, $C_0^{m_0(j)} = 0$ in $S^{(j)}$, then we have $f_0'' = \bar{p}_0$; or
  (b) if the SR-ARQ transmitter receives a NACK, i.e., given $C_0^{m_0(v)} = T_0$ in $\tilde{S}^{(v)}$, $C_0^{m_0(j)} = 0$ in $S^{(j)}$, then we have $f_0'' = p_0$.

(4) In all the other cases, $f_0'' = 0$.

Let us now collect all the $P_{i,j}^{+'}$ into the matrix $P^{+'} = \left(P_{i,j}^{+'} \right)$. Similarly, let us collect all the $P_{i,j}^{+''}$ into the matrix $P^{+''} = \left(P_{i,j}^{+''} \right)$. Therefore, the matrix P^+ can be expressed as $P^+ = P^{+'} \times P^{+''}$, which is the product of the matrices $P^{+'}$ and $P^{+''}$. Finally, the probability transition matrix $P = (P_{i,j})$ can be formed as $P = \left(P_{i,j} = P_{i,j}^+ + P_{i,j}^0 \right) = P^+ + P^0$, where $P^+ = \left(P_{i,j}^+ \right) = \left(P_{i,j}^{+'} \times P_{i,j}^{+''} \right)$ and $P^0 = \left(P_{i,j}^0 \right)$, respectively. With the above preparation, the steady-state throughput of the network coding node operated under the SR-ARQ can now be analyzed, as shown in Sect. 2.4.3.

2.4.3 Steady-State Throughput

As for the GBN-ARQ in Sect. 2.3.3, let $P_i(m)$ denote the probability that the state of node B is $S^{(i)}$ at time $t = mT$, where $1 \leq i \leq L$. Let $p(m) = [P_1(m), P_2(m), \ldots, P_l(m) \ldots, P_L(m)]^T$. Furthermore, let $\pi = \lim_{m \to \infty} p(m)$, where $\pi = [\pi_1, \pi_2, \cdots, \pi_L]^T$. Then, following Sect. 2.3.3, we have

$$\pi = P^T \pi \tag{2.13}$$

under the constraint of $\sum_{1 \leq i \leq L} \pi_i(m) = 1$.

According to the operation principles as detailed in Sect. 2.4, throughput is generated only when the state transitions of node B result in the non-zero transition probabilities of $\{P_{i,j}^+\}$, which are the non-zero elements in matrix P^+. Therefore, the steady-state throughput of the HISO network coding system can be expressed as

$$R_c = \sum_{i=1}^L \pi_i \sum_{j=1}^L P_{i,j}^+ \times \gamma_{i,j} \tag{2.14}$$

where the throughput R_c has the unit of the number of packets per time-chip. Furthermore, when assuming the same RTT of T, where T is the number of time-chips in one RTT, the throughput can also be express as $R_{RTT} = R_c T$, where R_{RTT} is steady-state throughput having the unit of the number of packets per RTT.

2.5 Performance Results

In this section, we provide some simulation results, in order to characterize the throughput performance of the HISO network coding system using the GBN-ARQ or SR-ARQ, as shown in Fig. 2.1a or b, and to justify our analytical results obtained in the previous sections. In our considered examples, we assume that the PER of links $l_0, l_1, \ldots, l_H$ is the same and equal to p_e. We also assume that the RTT of links $l_0, l_1, \ldots, l_H$ is the same and equal to T. Finally, we assume that $W_1 = W_2 = \ldots = W_H = W$. In our simulations, the throughput is normalized by the duration of RTT, and hence the normalized throughput at $t = mT$ is expressed as

$$R_{RTT}(m) = \frac{N(m)}{m}, \quad m = 1, 2, \ldots \tag{2.15}$$

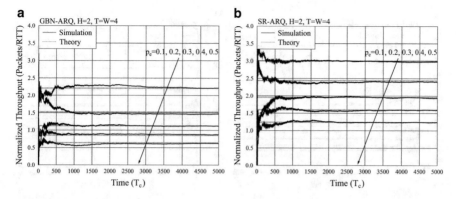

Fig. 2.2 Normalized throughput of the network coding system as shown in Fig. 2.1a or b. (**a**) GBN-ARQ. (**b**) SR-ARQ

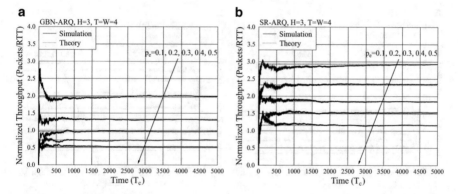

Fig. 2.3 Normalized throughput of the network coding system as shown in Fig. 2.1a or b. (**a**) GBN-ARQ. (**b**) SR-ARQ

where $N(m)$ represents the number of packets successfully transmitted from node B to node C during the period from $t = 0$ to the mth RTT.

Figures 2.2–2.5 depict the normalized throughput of the network coding system having $H = 2, 3, 4, 5$ incoming links, respectively. Each of the buffers is assumed to be able to store a maximum number of four packets. The length of the transmission window for each GBN-ARQ transmitter is W. Following the assumptions made throughout our analysis, we assumed that $T = W = 4$. In these figures, the corresponding steady-state throughput values evaluated from formula (2.9) for GBN-ARQ or (2.14) for SR-ARQ were depicted for the sake of comparison. From the results of Figs. 2.2–2.5, we can have the following observations. First, the throughput starts appearing from $R = 0$ at $t = 0$, and, finally, converges to the steady-state throughput as more packets are transmitted. As shown in Figs. 2.2–2.5, throughput obtained by simulations fluctuates around its corresponding steady-state throughput obtained from evaluation of (2.9) or (2.14), which is due to an

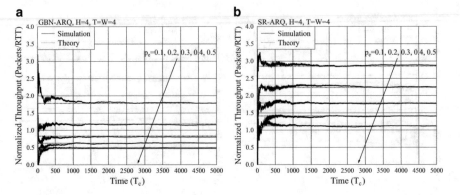

Fig. 2.4 Normalized throughput of the network coding system as shown in Fig. 2.1a or b. (**a**) GBN-ARQ. (**b**) SR-ARQ

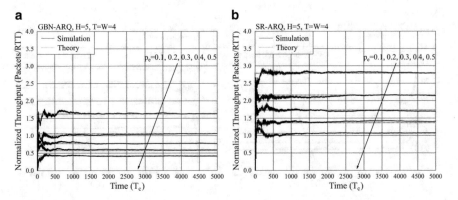

Fig. 2.5 Normalized throughput of the network coding system as shown in Fig. 2.1a or b. (**a**) GBN-ARQ. (**b**) SR-ARQ

insufficient number of packets transmitted. Second, the simulation results shown in Figs. 2.2–2.5 justify our analytical results obtained in Sect. 2.3. Therefore, the formulas obtained are effective for evaluation of the steady-state throughput of the HISO network coding systems employing GBN-ARQ or SR-ARQ. Third, as shown in Figs. 2.2–2.5, the normalized throughput decreases significantly, as the PER p_e increases. This is because, as p_e increases, more frequent re-transmissions are required. Fourth, when comparing the throughput results shown in Figs. 2.2–2.5, we can find that, for a given p_e, the throughput with a larger value of H is lower than that with a smaller value of H. In other words, the throughput decreases, as the number of incoming links of H to the coding node increases. The reason for this observation is obvious, as more incoming links to the coding node requires in average more waiting time to prepare the packets for forming a new coded packet, which will become more clear in the following figures. Finally, when comparing the

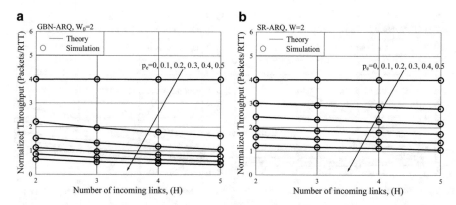

Fig. 2.6 Normalized throughputs of the network coding system as shown in Fig. 2.1a or b with $T = W = 4$ and $H = 2, 3, 4, 5$, respectively. (**a**) GBN-ARQ. (**b**) SR-ARQ

throughput achieved by the GBN-ARQ with that attained by the SR-ARQ, we can see that, for the same system settings, the SR-ARQ outperforms the GBN-ARQ.

In Fig. 2.6, we compare the normalized steady-state throughput against the PER, p_e, for the network coding node B, when it has $H = 2, 3, 4$, or 5 incoming links. Explicitly, at a given p_e, the normalized throughput decreases, when the network coding node has more incoming links. Again, this is because the chance for the network coding node to form a new coded packet and forward it to buffer b_0 becomes smaller, as a coded packet depends on correctly receiving more packets from the incoming links. From Fig. 2.6 we observe that the difference of the normalized throughput corresponding to different number of incoming links becomes smaller, as the PER increases. As p_e increases, the impact of the number of incoming links H on the achievable throughput becomes less. Furthermore, Fig. 2.5 shows that the steady-state throughput drops significantly as the p_e increases. Additionally, the SR-ARQ always achieves a higher throughput than the GBN-ARQ, provided that the PER is not zero.

Finally, in Fig. 2.7 we compare the normalized steady-state throughput against the PER, p_e, for the network coding node B, using GBN-ARQ or SR-ARQ, when the number of incoming link is $H = 2$ and the transmission window $W = 2, 4, 6, 8$, respectively. As seen in Fig. 2.7, for a given p_e, the throughput increases as the window size W increases, which is more declared when the SR-ARQ is employed. Furthermore, as the PER increases, the impact of W on the throughput becomes smaller.

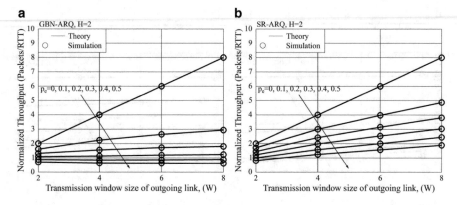

Fig. 2.7 Theoretical throughputs of the network coding node shown in Fig. 2.1a or b with $H = 2$ and $T = W = 2, 4, 6, 8$, respectively. (**a**) GBN-ARQ. (**b**) SR-ARQ

2.6 Conclusions

In this chapter, the steady-state throughput of the HISO network coding nodes assisted by the GBN-ARQ or SR-ARQ scheme has been investigated. Expressions for computing the steady-state throughput of the network coding node has been derived by assuming that each of the incoming links has one buffer unit for the GBN-ARQ or has some buffer units for the SR-ARQ to store the received packets, and that the outgoing link has a buffer to store the packets to be transmitted. The steady-state throughput performance of the GBN-ARQ or SR-ARQ-assisted network coding node has been investigated by both simulations and analytical approaches. It can be shown that the simulation results justify our analytical results derived. Furthermore, the performance results show that the throughput of a coding node decreases, as the number of incoming links attached to the coding node increases. This property implies that, in a network coding system, the embedded coding nodes may form the bottlenecks for information delivery, especially, when the number of incoming links is relatively high. The steady-state throughput of the network coding node increases, as the size of the transmission windows increases. However, the steady-state throughput drops significantly, as the PER increases. Additionally, our studies show that the SR-ARQ outperforms the GBN-ARQ in terms of the throughput performance.

References

1. R. Ahlswede, N. Cai, S.Y.R. Li, R.W. Yeung, Network information flow. IEEE Trans. Inf. Theory **46**(4), 1204–1216 (2000)
2. D.S. Lun, M. Medard, M. Effros, On coding for reliable communication over packet networks, in *Proceedings of the 42nd Annual Allerton Conference on Communication, Control, and Computing*, September/October 2004

3. T.Y. Chung, C.-C. Wang, Y.-M. Chen, Y.-H. Chang, PNECOS: a peer-to-peer network coding streaming system, in *IEEE International Conference on Sensor Networks, Ubiquitous and Trustworthy Computing, 2008. SUTC '08*, June 2008, pp. 379–384

4. C. Gkantsidis, P.R. Rodriguez, Network coding for large scale content distribution, in *Proceedings of the IEEE 24th Annual Joint Conference of the IEEE Computer and Communications Societies INFOCOM 2005*, vol. 4, March 13–17, 2005, pp. 2235–2245

5. S.Y.R. Li, R.W. Yeung, N. Cai, Linear network coding. IEEE Trans. Inf. Theory **49**(2), 371–381 (2003)

6. D.S. Lun, M. Medard, R. Koetter, M. Effros, Further results on coding for reliable communication over packet networks, in *Proceedings of the International Symposium on Information Theory '05*, 4–9 September 2005, pp. 1848–1852

7. C. Fragouli, D. Lun, M. Medard, P. Pakzad, On feedback for network coding, in *Proceedings of the 41st Annual Conference on Information Sciences and Systems 2007*, 14–16 March 2007, pp. 248–252

8. J.K. Sundararajan, D. Shah, M. Medard, ARQ for network coding, in *Proceedings of the IEEE International Symposium on Information Theory 2008*, 6–11 July 2008, pp. 1651–1655

9. J.K. Sundararajan, D. Shah, M. Médard, Online network coding for optimal throughput and delay—the two-receiver case, in *ACM Computing Research Repository*, vol. abs/0806.4264, 2008.

10. D. Nguyen, T. Tran, T. Nguyen, B. Bose, "Hybrid ARQ-random network coding for wireless media streaming," in *Proc. Second International Conference on Communications and Electronics 2008*, 4–6 June 2008, pp. 115–120.

11. P.A. Chou, Y. Wu, K. Jain, Practical network coding, in *Proceedings of the 41st Annual Allerton Conference on Communication, Control, and Computing*, 2003

12. J. Barros, R. Costa, D. Munaretto, J. Widmer, Effective delay control in online network coding, in *Proceedings of the INFOCOM 2009*, April 2009, pp. 208–216

13. J.K. Sundararajan, P. Sadeghi, M. Médard, A feedback-based adaptive broadcast coding scheme for reducing in-order delivery delay, in *Workshop on Network Coding, Theory, and Applications 2009*, June 2009, pp. 1–6

14. Y. Sagduyu, A. Ephremides, On broadcast stability of queue-based dynamic network coding over erasure channels. IEEE Trans. Inf. Theory **55**(12), 5463–5478 (2009)

15. Y. Qin, L.-L. Yang, Throughput comparison of automatic repeat request assisted butterfly networks, in 2010 7th International Symposium on Wireless Communication Systems (ISWCS), September 2010, pp. 581–585

16. Y. Qin, L.-L. Yang, Delay comparison of automatic repeat request assisted butterfly networks, in *2010 7th International Symposium on Wireless Communication Systems (ISWCS)*, September 2010, pp. 686–690

17. Y. Qin, L.-L. Yang, Throughput analysis of stop-and-wait automatic repeat request scheme for network coding nodes, in *2010 IEEE 71st Vehicular Technology Conference (VTC 2010-Spring)*, May 2010, pp. 1–5

18. Y. Qin, L.-L. Yang, Throughput analysis of general network coding nodes based on SW-ARQ transmission, in *2010 IEEE 72nd Vehicular Technology Conference Fall (VTC 2010-Fall)*, September 2010, pp. 1–5

19. Y. Qin, L.-L. Yang, Performance of general network coding nodes with stop-and-wait automatic repeat request transmission. IET Commun. **6**(15), 2465–2473 (2012)

20. Y. Qin, L.-L. Yang, Steady-state throughput analysis of network coding nodes employing stop-and-wait automatic repeat request. IEEE/ACM Trans. Networking **20**(5), 1402–1411 (2012)

21. Y. Qin, L.-L. Yang, Delay analysis of network coding nodes and butterfly network employing stop-and-wait automatic repeat request. IET Commun. **7**(5), 490–499 (2013)

22. S.M. Ross, *Introduction to Probability and Statistics for Engineers and Scientists*, 4th edn. (Academic, London 2009)

23. R.A. Horn, *Matrix Analysis* (Cambridge University Press, Cambridge, 1990)

24. D.P. Bertsekas, R.G. Gallager, *Data Networks* (Prentice Hall, Upper Saddle River, NJ, 1992)

Chapter 3
Network Coding at Network Layer in Multi-hop Wireless Networks

Yang Qin and Xiaoxiong Zhong

Abstract Network coding (NC) has been proved to be a breakthrough in the research area of wire networks and wireless networks, which can significantly increase network capacity, the throughput, robustness, and reliability of the networking systems. In this study, we present fundamental concepts related to the applications of the state-of-the-art NC at network layer in wireless networks. To begin with, we present a detailed investigation and comparison of the current classical routing protocols that incorporated NC and routing in wireless network. Then, we introduce several schemes of network coding with multicast at network layer. Last, we propose two opportunistic routing protocols based on NC technology for cognitive radio networks (CRNs).

3.1 Introduction

In recent years, network coding is relatively a new field of research. The idea was first introduced to satellite networks in [1], and then it was fully developed in 2000, when Ahlswede et al. published the seminar paper [2]. Through network coding, the multicast capacity of communication networks can be achieved. Later, Li et al. [3] showed that liner codes are sufficient for multicast traffic to achieve the maximum capacity bounds. At the same time, Koetter and Medard [4] proposed an algebraic approach and showed that coding and decoding can be done in polynomial time. In 2006, Ho et al. [5] presented the concept of random linear network coding, which makes network coding more practical, especially in distributed networks including wireless networks.

Y. Qin (✉)
Computer Science Department, Shenzhen Graduate School, Harbin Insititute of Technology, Xili, Shenzhen, Guangdong, China
e-mail: yqinsg@gmail.com

X. Zhong
School of Computer Science and Technology, Harbin Institute of Technology Shenzhen Graduate School, Shenzhen 518055, People's Republic of China

School of Computer Science and Engineering, Guilin University of Electronic Technology, Guilin 541004, People's Republic of China
e-mail: xixzhong@gmail.com

© Springer International Publishing Switzerland 2016   59
Y. Qin (ed.), *Network Coding at Different Layers In Wireless Networks*,
DOI 10.1007/978-3-319-29770-5_3

Network coding has been widely considered as an effective approach to improve the performance of wireless networks. In the last few years, many researchers have put their efforts to develop viable network coding techniques in wireless networks [6, 7]. Recently, a great deal of attention has been focused on dealing with practical issues and developing implementable protocols with network coding [8–10]. Generally speaking, network coding techniques in wireless networks can be divided into two categories, *intersession network coding* and *intra-session network coding*, also known as coding-aware and coding-based. In the former, coding is operated on packets from different flows; while in the latter; coding is done over the packets belonging to the same flow. These two network coding techniques can increase the overall throughput of wireless networks from different aspects as we will explain next.

Nevertheless, the idea behind network coding is extraordinarily simple. As stated in [11], network coding is no more than performing coding on the contents of packets, by arbitrarily mapping the contents of packets rather than just duplicating and forwarding the packet which is typically allowed in conventional, store-and-forward network architectures.

When network coding is considered for a single session, for example, unicast session from one source to one destination or multicast session from one source to multiple destinations, it is called as intra-session NC. In this type of coding, only the information belonging to the same session is coded together. Each node can form coded packet as random linear combinations of the input packets, and each destination node can decode once it receives enough independent linear combinations of the source packets. Intersession network coding usually will select the packets from different sessions to code together. This technology could be deployed at router or relay nodes where traffic will aggregated to be forwarded. Recently, some research presented a scheme to combine these two methods called I2NC [12].

So far, a few routing protocols which exploit network coding technology in wireless networks have been proposed. In this chapter, the representative schemes are introduced. Then, two routing protocol-based network codings for cognitive radio networks (CRNs) are presented. In the meantime, typical schemes of network coding with multicast at network layer have been introduced in this chapter as multicast is also an important function at network layer.

The rest of this chapter is organized as follows. Section 3.2 describes some classical routing + NC schemes in wireless networks. Two routing protocol-based network codings for CRNs are proposed in Sect. 3.3. Section 3.4 concludes this chapter.

3.2 Applications of Network Coding at Network Layers in Wireless Networks

In this section, we will address some classical routing scheme-based network coding at network layer. In addition, some recent research results of applications of network coding with multicast at network layer will also be introduced.

3.2.1   Classical Routing Scheme-Based NC in Wireless Networks

COPE [8] is the first practical and representative coding-aware routing for wireless networks, which can largely increase network throughput in wireless networks. In COPE, it incorporates three main techniques:

1. Opportunistic listening (sets the nodes in promiscuous mode, makes them snoop on all communications over the wireless medium and store the overheard packets for a limited period)
2. Opportunistic coding (under the circumstance that the intended next hop can successfully decode the coding packets, it maximizes the number of native packets delivered in a single transmission)
3. Learning neighbor state (each node announces to its neighbors the packets it stores in reception reports and exploits a mechanism to guess whether a neighbor has a particular packet)

In COPE, it mainly investigates coding opportunities for two classical topologies (chain topology and "X" topology as shown in Fig. 3.1). Its coding gain depends on the existence of coding opportunities, which themselves depend on the traffic patterns. However, COPE has two drawbacks: (1) the coding topologies are strictly limited two-hop coding pattern and (2) it separates the process of route discovery and coding opportunity discovery.

Figure 3.1 illustrates the basic idea of network coding for data transmission in wireless networks. Figure 3.1a shows the chain topology, and Fig. 3.1b shows the X topology. In both cases, node R is the common relay node for two flows. When node R receives packets P1 and P2, it encodes two packets and then broadcasts P1 ⊕ P2. Upon receiving the coded packet, nodes C and D can decode the coded packet.

Both coding structures shown in Fig. 3.1 are restricted within a two-hop region. In these cases, the coding opportunities occur within the region that includes only the relay node and its one-hop predecessors and successors.

To address the abovementioned limitations of COPE, Le et al. [10] proposed distributed coding-aware routing (DCAR) protocol for the k-hop ($k > 2$) coding model in wireless networks. They gave a coding-aware path discovery mechanism-

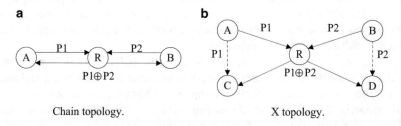

Chain topology.   X topology.

Fig. 3.1 Examples of data transmission with network coding. (**a**) Chain topology. (**b**) X topology

Fig. 3.2 Single coding node
in the k-hop coding model

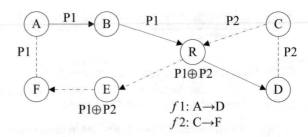

coding + routing and first stated the necessary and sufficient conditions of network coding to facilitate the coding opportunity discovery.

The *k*-hop coding model is illustrated in Fig. 3.2. Suppose that two flows A → D and C → F intersect at R and D, F can overhear the transmission from A and C. Although E cannot decode the coded packet P1 ⊕ P2, it only forwards the coded packet to F. F can decode the coded packet, because it has overheard packet P1.

The DCAR has two characteristics: it has the ability to discover available paths and concurrently detect coding opportunities on the entire paths; it has the ability to find paths that have more coding opportunities and ultimately higher throughput that eliminates the two-hop coding limitations in COPE.

In DCAR, the coding condition may be invalid when there are multiple intersecting nodes along a path, so Guo et al. [13] proposed general network coding conditions for multi-flow wireless networks-free-ride-oriented routing (FORM), which include a coding flow selection procedure.

In FORM, the coding conditions have two important properties:

Availability: If the packets are encoded at the coding node, the destination can obtain the corresponding native packet, either by decoding the encoded packet itself or through some decoding-capable nodes that have retrieved the native packet for it.
Compatibility: It guarantees that the encoding function of the coding nodes introduces no negative impact on other flows, which are overlooked by some related work.

All abovementioned coding-aware routing schemes are based on traditional routing, which always select one "best" path for data transmission. Opportunistic routing [14] takes advantage of the broadcast nature of wireless medium and explores forwarding capacity of all candidate nodes that overhear packets, which can significantly improve the performance of wireless networks. In recent years, there are lots of research works that combined network coding and opportunistic routing (CNCOR) for wireless networks.

Coding-aware opportunistic routing mechanism (CORE) [15] is a classical CNCOR scheme which combines hop-by-hop opportunistic forwarding and localized interflow network coding for enhancing the throughput performance of wireless mesh networks. CORE consists of four major components: forwarder set selection, coding opportunity calculation, forwarder selection, and priority-based forwarding. In CORE, the candidates received the packet collaborate to choose the next-hop nodes with the most coding opportunities in a localized fashion. It attempts to maximize coding gain in a single transmission by network coding technology.

Multi-rate opportunistic routing-aware network coding (MRORNC) [16] is proposed for multi-rate wireless networks, which jointly determines the coded packet, potential next hops, and transmission rate using a novel metric.

In each transmission opportunity, MRORNC jointly chooses the native packets of the coded packet, the candidate next hops of each native packet, and the transmission rate so as to maximize the forwarding efficiency.

XCOR (network coding with opportunistic routing) [17] is designed to facilitate multiple flows, which integrate intersession network coding for data transmission in wireless networks.

In XCOR, each relay node checks packets from different flows and codes them only if the utility obtained after network coding is larger than that of without network coding.

CoAOR [18] takes into account the number of flows considered in an encoded packet, the link quality, and the number of nodes that are able to decode an encoded packet. The forwarder set selection in CoAOR is different from that in CORE. In CORE, a forwarder node only considers the forwarder candidates of a particular primary packet and ignores the potential next-hop forwarders of other primary packets in the same encoded packet. Therefore, a selected forwarder set may only be suitable for a particular primary packet, not the encoded packet. In CoAOR, a coding-aware forwarder set selection method is used, which can significantly improve the performance in terms of the network throughput and the average processing time for forwarding a packet as compared with the CORE.

In OR, the coordination among forwarding candidates is an important issue, which is used to avoid duplication transmissions. Integrating intra-session network coding and OR will be an efficient solution for the coordination problem. MORE (MAC-independent opportunistic routing and encoding) [9] is a first protocol that exploits RLNC for OR, which can reduce the probability of useless transmissions and solve the coordination of the forwarding candidates in OR.

In MORE, intermediate nodes perform NC operation on packets of the same session. In intra-session NC, the source node divides the file into batches of k packets, $P_1, \ldots, P_k$. These k uncoded packets are called native packets. The source node creates a random linear combination of the k native packets in the current batch in the form $P' = \sum_{x=1}^{k} a_x P_x$, where a_x, $\forall x$ is a random coefficient chosen over a finite field of a large enough size, typically 2^8. We call $\overrightarrow{a} = (a_1, \ldots, a_k)$ the packet's code vector. A coded packet is $P' = \sum_{x=1}^{k} a_x P_x$. The source node keeps sending coded packets from the current batch until the batch is acknowledged by the destination node. And then, it moves to the next batch. Upon receiving a coded packet, the intermediate node checks to see if the coded packet is linearly independent from the packets the node has previously received from this batch. If so, it stores the coded packet; otherwise, it drops the packet. Also, it linearly combines the coded packet to create more coded packets as follows: $P'' = \sum_{y=1}^{k} \omega_y P'_x$, where ω_ys are random numbers. Hence, the coded packet can be expressed as $P'' = \sum_{y=1}^{k} \left(\omega_y \sum_{x=1}^{k} a_x P_x \right)$. When the destination node receives any k linearly independent packets, this means that it can decode all of the packets of the batch. The destination node decodes the whole batch using simple matrix inversion:

$$\begin{pmatrix} P_1 \\ \vdots \\ P_k \end{pmatrix} = \begin{pmatrix} a_{11} \ldots a_{1k} \\ \vdots \ddots \vdots \\ a_{k1} \cdots a_{kk} \end{pmatrix}^{-1} \begin{pmatrix} P'_1 \\ \vdots \\ P'_k \end{pmatrix} \tag{3.1}$$

where P_x is a native packet and P'_x is a coded packet whose code vectors is $\overrightarrow{a}_x = (a_{x1}, \ldots, a_{xk})$. Therefore, it sends an ACK to the source and notifies it to stop sending from the current batch and to move to the next batch.

Through intra-session NC, MORE needs no special coordination mechanism among forwarding candidates, which can avoid the duplication transmissions. However, in MORE, there still exist some useless packet transmissions, which will influence the resource consumption and performance of OR.

CodeOR [19]: In MORE, the source node divides the file into multiple batches and encodes packets of the same batch. The source node only transmits the packets of the same batch until receiving the acknowledgement for the batch, which will degrade performance of MORE as the network size scales up. In MORE, too much batches transmitting concurrently may lead to congestion, while only one batch transmitting will degrade the performance. To solve this problem, Lin et al. introduced the sliding window mechanism of TCP flow control into MORE and proposed a novel opportunistic routing, CodeOR (coding in opportunistic routing). In CodeOR, each node has a sliding window to limit that only packets of batches in the window could be transmitted in the whole network. In addition, CodeOR uses end-to-end acknowledgement (E-ACK) and hop-by-hop acknowledgement (H-ACK). E-ACK is transmitted to the source node only when batch i and all batches before i have been decoded at the destination node to notify that the source node could move the sending window to $i + 1$. A node exploits H-ACK to inform that the upstream node can start to transmit new batches and avoid redundant transmission.

Through transmitting multiple batches simultaneously, the CodeOR significantly improves the performance of MORE and is particularly suitable for large-scale networks.

CCACK [20]: The prior schemes leverage credit mechanism using measurements of offline loss rate to control the transmission of coded packets, which will lead to the performance mainly depending on the accuracy and freshness of the loss rate measurements. In 2010, CCACK (cumulative coded acknowledgment), an improved version of MORE, is a proposed approach oblivious to the loss rate which allows nodes to acknowledge network-coded traffic to their upstream nodes in a simple way and with practically zero overhead.

Knowing which packet has been received by the forwarding candidates, upstream nodes will send only the packets innovative to forwarding candidates and avoid unnecessary transmission.

3.2.2 Network Coding for Multicast at Network Layer

Multicast is an important function at network layer. There are several successful deployments of network coding with multicast to improve the multicast efficiency. Li et al. addressed the problem of rate control in the context of network coding-based multicast flows with elastic rate demand for wired networks with given coding subgraphs or without given coding subgraphs in [21]. The proposed multicast scheduling protocol takes some important factors, which include power control, relay assignment, buffer management, dynamic spectrum access, and fairness, into consideration. And in this new framework, base station multicast data to a subnet of secondary users by tuning the power while secondary users can opportunistically use locally idle primary channels to perform cooperative transmissions concurrently. And intra-network coding is used to reduce overhead and increase reliability during the transmission.

In order to implement the multicast scheduling protocol, authors develop two kinds of algorithms to find the optimal power control on BS and the most efficient cooperative communication schedule: one greedy protocol for the centralized optimization and the other online cooperative protocol for distribute manners. They are designed based on centralized greedy optimization and stochastic Lyapunov optimization by using a sound theoretical foundation separately.

Junchen Li et al. addressed the problem of rate control in the context of network coding-based multicast flows with elastic rate demand for wired networks with given coding subgraphs or without given coding subgraphs [22].

The one with given coding subgraphs is an approach which uses uncapacitated subgraphs that only specify which links are used by a session but not the amount of data sent on each link. The coding subgraphs are chosen based on some common cost criteria which are independent of flow rates. The advantage of this algorithm is much less complex, but it gets a lower throughput than the one without given coding subgraphs.

The other one without given coding subgraphs is desired to make dynamic routing and coding decision based on queue length gradients instead of explicitly finding coding subgraphs. Numerical results show that multicasting algorithm without given coding subgraphs gets a higher throughput since the capacity region for it is a superset of the capacity region with coding subgraphs.

The two algorithms come from the solution of two optimization problems P1 and P2 which are convex problems with strong duality; distributed algorithms can be derived by formulating and solving corresponding Lagrange dual problems. And both algorithms can be implemented in a distributed manner, and the rate control part works at transport layer to adjust source rates, and network coding works at network layer to carry out random network coding:

$$\textbf{P1} : \max_{x_r^m, y_l^m} \sum_m U_m(x^m)$$

$$\text{Subject to } H_{lr}^m x_r^m \leq y_l^r, \forall r \in R_m, \forall m \in M$$

$$\sum_m y_l^m \leq c_l, \forall l \in L$$

$$\textbf{P2} : \max_{x,g,f} \sum_m U_m(x^m)$$

$$\text{Subject to } \sum_{j:(i,j) \in L} g_{i,j}^{md} - \sum_{j:(j,i) \in L} g_{j,i}^{md} = x_i^m, i \neq d, \forall d, m$$

$$g_{i,j}^{md} \leq f_{i,j}^m, \forall d, m$$

$$\sum_m f_{i,j}^m \leq c_{i,j}, \forall (i,j) \in L$$

where x^m is the rate of session m and $U_m(x^m)$ is its utility. P1 tries to find a transmit rate x^m which can maximize the utility, while P2 tries to choose a source rate x^m, information rates $g_{i,j}^{m,d}$, and link capacity allocation $f_{i,j}^{m,d}$ to maximize the aggregate utility.

Wu et al. proposed a new network coding-based retransmission named CoRET for mobile communication networks [23]. CoRET uses a Hamming distance-based packet selection algorithm instead of the random selection. To support packet selection, the base station has to keep a table which could specify status of all the packets sent for receivers. Thus, each packet p_i uses a vector $V_{pi}(k)$ which has k elements to represent its receiver status at k receivers and use 1 for successful and 0 for failed receiver. The algorithm always selects the two packets which have the largest Hamming distance to encode with each other, and the Hamming distance can be calculated by the following equation:

$$\text{Diff}(i,j) = \sum_{k=1}^{n} |V_{P_i}(k) - V_{P_j}(k)| \tag{3.2}$$

In this formula, n represents the number of packets. P_i, P_j represent packet i and j. And V_{P_i} is the vector of the coefficient of packet i on multicast receivers.

(a) Rate control.

　　　To support CoRET, the base station must introduce a request indicator bit (RIBs) in its signaling format to indicate the state of each packet. If the packet has been retransmitted, its RIB is set to 1; otherwise, its RIB is set to 0.

(b) Optimal encoding number.

In this part, a mathematical model is developed to describe the relationship among the encoding number, the retransmission efficiency, the retransmission, the reliability, and the retransmission overhead. Based on this model, the problem to find the optimal encoding number k can be formulated as an optimization problem with the objective

$$\text{MAX}\left[\frac{N_{\max}(k)}{N_{\text{ctrl}}(k)}\right]$$

Subject to
$k \geq 0$ and $k \in N$
$0 \leq r_i \leq R$ and $i \in N$
$0 \prec P_i^{r_i}(k) \prec 1$ and $0 \leq m_i^{r_i} \leq M$

where R is a predetermined number of transmission times. $N_{\text{ctrl}}(k)$ denotes the average number of control bits introduced for retransmitting an encoded packet in a transmission cycle. And $N_{\text{ctrl}}(k)$ can be calculated by the following:

$$N_{\text{ctrl}}(k) = \lim_{T\to\infty} \frac{8(k+1)\sum_{l=1}^{T} S_l(k)}{T}\text{bits} \tag{3.3}$$

$N_{\max}(k)$ denotes the maximum sequence number among all transmitted packets after T transmission cycles. And it can be formulated as the following:

$$N_{\max}(k) = \lim_{T\to\infty} \frac{\sum_{l=1}^{T}[n_s - S_l(k)]}{T} \tag{3.4}$$

In Eqs. (3.3) and (3.4), $S_l(k)$ is the number of the slots needed for retransmission in the l transmission cycle and $S_0(k) = 0$. $S_l(k)$ can be calculated by Eqs. (3.5) and (3.6), and $N_l(k)$ is the expected number of original packets retransmitted in the lth transmission cycle.

$$S_l(k) = \left\lceil \frac{N_l(k)}{k} \right\rceil \tag{3.5}$$

$$N_l(k) = S_{l-1}(k)\left[k - \sum_{i=1}^{k}(1-p_e)p_i^{r_i}(k)^{m_i^{r_i}(k)}\right] \tag{3.6}$$

This scheme can be extended to the situation that base station selects more than two packets to encode.

In the context of Hamming distance-based packet selection algorithm, Hao Wu also tries to find an optimal encoding number to achieve the best performance. And the numerical results show that a smaller encoding number is preferred when the transmission channel is reliable and a larger encoding number can achieve a better performance when the channel becomes unreliable.

Lu et al. designs a dynamic programming algorithm which is based on Markov decision process to solve the minimal delay transmission schedule in multi-rate relay-assisted wireless networks [24]. The proposed Markov decision process-based dynamic programming algorithm is able to find optimal retransmission strategy for any possible system state, and a minimal retransmission delay can be achieved with proposed strategy.

In order to tackle this, Lu designs two lightweight heuristic algorithms which allow the trade-off between solution quality and computational complexity. One is the estimated delay-based algorithm and the other is a greedy heuristics algorithm. The estimated delay-based algorithm is based on the fact that packet error rate is much smaller than 0.5 in practice so that only maintaining the transitions with high probabilities while neglecting others reduces the computational complexity. The greedy algorithm tries to select the strategy that could cover as many as SSs while consuming as little time as possible since their goal is to minimize the time required for all SSs to cumulatively receive N innovative packets for retransmission. The greedy algorithm makes decision based on current situation instead of determining the long-term delay prediction so that it gets a good performance.

It is concluded that the greedy retransmission scheme is the most cost-effective for multi-rate multicast in NC-enabled WRNs if RSs can cooperatively forward packets; otherwise the estimated delay-based scheme will be better.

Li et al. presents a reliable multicast protocol CodePipe in lossy wireless networks, with advanced performance in terms of throughput, energy efficiency, and fairness in [25]. CodePipe is built on opportunistic routing and random linear network coding. For the opportunistic routing part, CodePipe simplified the transmission coordination between nodes; while in network coding part, CodePipe tries to exploit both intra-batch and inter-batch coding opportunities to improve the multicast throughput. CodePipe have alleviated MORE's "stop-and-wait" nature between consecutive batches, to allow a siding window among the batches.

The proposed multicast protocol includes four main parts: linear programming-based opportunistic routing structure, opportunistic feeding, fast batch moving (similar to moving windows network coding manner), and inter-batch coding.

The proposed CodePipe comes from the solution of a linear programming formulation which tries to maximize the multicast throughput λ as follows:

Max λ subject:

$$\sum_{i:(s,i)\in E} f_{si}^d - \sum_{i:(i,s)\in E} f_{is}^d = \lambda, \forall d \in D_u$$

$$\sum_{i:(i,j)\in E} f_{ij}^d - \sum_{i:(j,i)\in E} f_{ji}^d = 0, \forall j \in V - \{s,d\}, \forall d \in D_u$$

$$\sum_{i:(d,i)\in E} f_{di}^d - \sum_{i:(i,d)\in} f_{id}^d = -\lambda, \forall d \in D_u$$

$$f_{ij}^d \leq c_{ij}t_i, \forall d \in D_u, \forall\, (i,j) \in E$$

$$f_{iJ_i}^d \leq c_{iJ_i}t_i, \forall d \in D_u, \forall i \in V$$

$$t_i + \sum_{j\in J_i} t_j \leq 1, \forall i \in V$$

where (i, J_i) is a hyper link, $J_i = \{j|\, (i,j) \in E\}$ is the receiver set, $f_{iJ_i}^d$ is the flow on hyper link (i, J_i), and t_i is the relative share of time allocated to transmitter i.

LP-based opportunistic routing structure defines a balanced opportunistic routing structure by taking opportunistic forwarding and contention probability into consideration. Opportunistic feeding could determine a destination whether it can act as a cooperative source to help forward the reception of current batch to other destination or not. Fast moving is also developed to accelerate the start of new batch if all cooperative sources have the ability to decode the current batch.

3.3 Two Routing Protocols with Network Coding in Multi-hop CRNs

In this section, we propose two routing protocol-combined network coding and routing in multi-hop cognitive radio networks: one is an intra-session network coding-based scheme and the other one is a network coding-aware scheme.

3.3.1 CANCOR: Intra-session Network Coding-Based Routing for CRNs

The cognitive radio principle has introduced the idea to exploit spectrum holes (i.e., bands) which result from the proven underutilization of the electromagnetic spectrum by modern wireless communication and broadcasting technologies [26]. Cognitive radio networks (CRNs) have emerged as a prominent solution to improve the efficiency of spectrum usage and network capacity. In CRNs, secondary users (SUs) can exploit channels when the primary users (PUs) currently do not occupy the channels. The set of available channels for SUs is instable, varying over time and locations, which mainly depends on the PU's behavior. Thus, it is difficult to create and maintain the multi-hop paths among SUs through determining both the relay nodes and the available channels to be used on each link of the paths.

In OR, instead of first determining the next hop and then sending the packet to it, a node with OR broadcasts the packet so that all neighbors of the node have the chance to hear it and assist in forwarding. OR provides significant throughput gains compared to traditional routing. In CRNs, it is hard to maintain a routing table due to dynamic spectrum access. The predetermined end-to-end routing does not suit for CRNs. Since opportunistic routing does not need prior setup of the route, it is more suitable for CRNs with dynamic changes of channel availability depending on the PU's behavior. However, it is not easy to extend the previous studies of OR in multi-hop CRNs. On the one hand, in CRNs, SU can only opportunistically explore the spectrum holes for communications. These spectrum holes are usually uncertain. On the other hand, the sender and forwarder should have a common available channel for communications. How to detect and choose a good candidate is not trivial for multi-hop CRNs.

The effects of opportunistic routing on the performance of CRNs have been investigated in [11]. In 2008, Pan et al. [27] proposed a novel cost criterion for OR in CRNs, which leverages the unlicensed CR links to prioritize the candidate nodes and optimally selects the forwarder. In this scheme, the network layer selects multiple next-hop SUs, and the link layer chooses one of them to be the actual next hop. The candidate next hops are prioritized based on their respective links' packet delivery rate, which in turn is affected by the PU activities. At the same time, Khalife et al. [28] introduced a novel probabilistic metric toward selecting the best path to the destination in terms of the spectrum/channel availability capacity. Considering the spectrum availability time, Badarneh et al. [29] gave a novel routing metric that jointly considers the spectrum availability of idle channels and the required CR transmission time over those channels. This metric aims at maximizing the probability of success (PoS) for a given CR transmission, which consequently improves network throughput. Lin et al. [30] proposed a spectrum-aware opportunistic routing for single-channel CRNs that mainly considers the fading characteristics of highly dynamic wireless channels. The routing metric takes into account transmission, queuing, and link-access delay for a given packet size in order to provide guarantee for end-to-end throughput requirement. Taking heterogeneous channel occupancy patterns into account, Liu et al. [31] introduced opportunistic routing into the CRNs where the statistical channel usage and the physical capacity in the wireless channels are exploited in the routing decision. Liu et al. [32] further discussed how to extend OR in multichannel CRNs based on a new routing metric, referred to as cognitive transport throughput (CTT), which could capture the potential relay gain of each relay candidate. The locally calculated CTT values of the links (based on the local channel usage statistics) are the basis for selecting the next-hop relay with the highest forwarding gain in the opportunistic cognitive routing (OCR) protocol over multi-hop CRNs.

However, none of the above schemes systematically combines the channel assignment with OR to model CRNs. The number of candidate forwarders and the performance of OR will decrease, if using existing channel assignment algorithms for multichannel multi-radio (MCMR) OR. A Workload-Aware Channel Assignment algorithm (WACA) for OR is designed in [33]. WACA identifies the nodes with high workloads in a flow as bottlenecks and tries to assign channels to these nodes with high priority. WACA is the first static channel assignment for OR. However, it deals with channel assignment for single flow. Assuming that the number of radios and the number of channels are equal, a simple channel assignment for opportunistic routing (SCAOR) is proposed in [34]. It selects a channel for each flow. SCAOR is for multiple flows, but assumes that the number of radios and the number of channels are equal in SCAOR.

There are some routing strategies [35, 36] which integrate channel assignment in their routing schemes. However, their schemes do not suit for OR in multichannel CRNs. On the one hand, OR improves throughput by making a certain downstream nodes operate on the same channel as the sender. On the other hand, the multichannel routing boosts throughput by distributing node/radios onto different channels, such that it can transmit simultaneously. But it will decrease the opportunity of a packet being heard by the downstream nodes. Hence, how to design an opportunistic routing with considering channel assignment is not a trivial task. However, the abovementioned channel assignment schemes are not a feasible solution for OR in MCMR CRNs due to channel uncertainty of SUs. In this paper, we combine channel assignment and opportunistic routing for maximizing the aggregate throughput of SUs.

In the following, we investigate the joint opportunistic routing and channel assignment problem in multichannel multi-radio (MCMR) cognitive radio networks (CRNs) for improving the aggregate throughput of the SUs. We first present the nonlinear programming optimization model for this joint problem, taking into account the feature of CRNs-channel uncertainty. In order to reduce the computational complexity of the problem, we present a heuristic algorithm to select forwarding candidates and assign channels in CRNs, including candidate selection algorithm, considering the queue state of a node and expected transmission count (*ETX*), and channel assignment algorithm, taking into account the transmission time and the available time of a given channel. Our simulation results show that the proposed scheme channel assignment and network-coded opportunistic routing (CANCOR) performs better than the traditional routing and classical opportunistic routing in which channel assignment strategy is employed.

3.3.1.1   System Model

We summarize the notations used in this paper in Table 3.1.

We model the CRNs as a directed graph denoted by $G = (V, E)$, where V is the set of N SUs and E is the set of links connecting any pair of nodes. The source node is denoted as S, and the destination is denoted as D. We consider time-slotted CRNs with K licensed orthogonal channels belonging to an interweave model [37]. The usage pattern of PUs over a given channel k that is available for SU transmissions follows an independent ON/OFF state model [38]. We assume that both ON and OFF periods are exponentially distributed with mean rate λ_p and μ_p. There are N SUs and M PUs in this CRN. Each node is equipped with the same number of radios R in half-duplex model. Each SU is capable of sensing the locally available channels and has the capability of channel changing at packet level for data transmission.

Table 3.1 Summary of key notations

Symbol	Meaning
$G = (V, E)$	The CRN topology graph
S	The source node
D	The destination node
ψ_i^{k+}	The set of node i's in-edge on the channel k
ψ_i^{k-}	The set of node i's out-edge on the channel k
ρ_{ij}^k	The loss rate of link e_{ij} ($e_{ij} \in E$) on the channel k
$\theta_i^k(t)$	The probability that node i can use the channel k on time slot t
P_i^k	The amount of packets that node i has sent on channel k
$\mu_{ij}^k(t)$	The probability of e_{ij} that transmits data packets using the channel k on time slot t
f_{ij}^k	The number of data packets that e_{ij} transmits on channel k
B	The maximum transmission rate on a channel
R	The number of radios
$Q_i(T)$	The queue length of i at time T

In CRNs, the SU's transmission range is d_s and the interference range is d_I. Let d_{ij} denote the distance between node i and node j. If $d_{ij} < d_s$, we say nodes i and j are neighbors. Node i and node j can communicate with each other if they are neighbors, and they are operating on the same channel. In OR, each node, i, has multiple candidate forwarders, denoted as CFS_i. For any two nodes, i and j, $i < j$ indicates that node i is closer to the destination node than node j, or in other words, i has a smaller ETX (expected transmission count) [39] than j.

3.3.1.2 Problem Formulation

In this section, we formulate the problem of joint channel assignment and OR as a nonlinear programming problem.

Let $h_{ij}^k \in \{0, 1\}$ denote whether node i and node j can communicate with each other through channel k. If $h_{ij}^k = 1$, it means that nodes i and j can communicate with each other and $h_{ij}^k = 0$, vice versa.

We adopt the protocol interference model [40]. If $d_{uj} \leq d_s$, it means that node j is in u's transmission range. When nodes i and u simultaneously transmit data packets, the transmission of i to j will interfere with the transmission of u to v in time slot t. Similarly, when $d_{iv} \leq d_s$, the transmission of node i to node j will interfere with the transmission of nodes u to v in time slot t. Thus, we can calculate the interference link set I_{ij} of link e_{ij}, $I_{ij} = \{< u, v > | < u, v > \in E, d_{uj} \leq d_s \text{ or } d_{iv} \leq d_s\}$, and have

$$\mu_{ij}^k(t) + \mu_{uv}^k(t) \leq 1, < u, v > \in I_{ij} \tag{3.7}$$

where $u_{ij}^k(t)$ is the probability of link e_{ij} that transmits data packets using the channel k in time slot t. In CRNs, $u_{ij}^k(t)$ is affected by the PU's activity. If two links are concurrently usable at the same channel in time slot t, they should either share the same transmitter or not interfere with each other. Hence, we can obtain

$$\sum_{<m,n>\in I_{ij}} \mu_{mn}^k(t) \leq 1, < m, n >\in I_{ij}, < i, j >\in E \tag{3.8}$$

Channel k can be allocated to link e_{ij} in time slot t only when channel k is available. Thus, we have

$$\mu_{ij}^k(t) \leq h_{ij}^k, < i, j >\in E \tag{3.9}$$

For each node i, it can participate in at most R simultaneous communications in any given time T (T includes some mini-slots t). This can be formally represented by

$$\begin{cases} \mu_{ij}^k(t) \leq \theta_i^k(t), < i, j >\in \psi_i^{k-} \\ \mu_{gi}^k(t) \leq \theta_i^k(t), < g, i >\in \psi_i^{k+} \\ \sum_k \theta_i^k(t) \leq R \\ 0 \leq \theta_i^k(t) \leq 1 \end{cases} \tag{3.10}$$

where $\theta_i^k(t)$ is the probability that node i can use the channel k in time slot t, ψ_i^{k+} is the set of node i's in-edge on channel k, and ψ_i^{k-} is the set of node i's out-edges on channel k.

In MORE, the candidate forwarder is selected according to ETX. However, since in the real world the buffer of a node is limited, it is reasonable to consider the buffer size in the packet forwarding scheme. Thus, we should take buffer size constraint into account to select forwarding candidate. During time slot T, node i sends P_{ik} packets on channel k. Then the queue length of i at time $(T + 1)$, $Q_i(T + 1)$, can be expressed as

$$Q_i(T + 1) = Q_i(T) - \sum_k P_i^k + \left(\sum_k \sum_t \sum_{<g,i>\in \psi_i^{k+}} f_{gi}^k \times \left(1 - \rho_{gi}^k\right) \times \mu_{gi}^k(t) \right) \tag{3.11}$$

where ρ_{gi}^k is the loss rate of link e_{gi} on channel k, f_{gi}^k is the number of data packets that e_{gi} transmits on channel k, P_i^k is the number of packets that node i has sent on channel k during time slot $T + 1$, and $\left(\sum_k \sum_t \sum_{<g,i>\in \psi_i^{k+}} f_{gi}^k \times \left(1 - \rho_{gi}^k\right) \times \mu_{gi}^k(t) \right)$ is the amount of receiving packets of node i during time slot $T + 1$.

For a given time T, the incoming packets of node i are the same as the outgoing packets of node i on channel k to keep traffic balance. Also, the total number of data packets that e_{ji} transmits on channel k are not exceeding the maximum transmission rate of the channel. Therefore, we have

$$\begin{cases} 0 \le f_{ij}^k \le \sum_t \mu_{ij}^k(t) \times B, <i,j> \in E \\ \sum_k \sum_{<g,i> \in \psi_i^{k+}} f_{gi}^k \times \left(1 - \rho_{gi}^k\right) \times \alpha_{gi}^k = \sum_k P_i^k + Q_i(T), i, g \in V, i \ne S \end{cases}$$

(3.12)

where B is the maximum transmission rate on a channel and α_{ij}^k is the forwarding probability that node j forwards the packet received from node i over channel k to j's next hop.

Generally, the channel availability is heterogeneous in CRNs due to PU's activity. So, in our scheme, in each intermediate node, we attach the forwarding probability α_{ij}^k to each data packet, which can reduce duplicate transmission. In the following, we give the **Algorithm 1** to calculate α_{im}^k.

The probability that only node m has received the packet is

$$\beta_{im}^k(1) = \rho_{i1}^k \rho_{i2}^k \rho_{im-1}^k \left(1 - \rho_{im}^k\right) \rho_{im+1}^k \cdots \rho_{il}^k$$

(3.13)

where ρ_{i1}^k corresponds to node 1 whose ETX_1 is the least in CFS_i^k, ρ_{i2}^k is node 2 whose ETX_2 is in the second place, and so on, and l is the number of candidates in CFS_i^k. All nodes, from 1 to l, are ordered by their *ETX*.

The probability that node m and at least one node in CFS_i^k $(m+1, \ldots l)$ have received the packet is

$$\beta_{im}^k(2) = \rho_{i1}^k \rho_{i2}^k \rho_{im-1}^k \left(1 - \rho_{im}^k\right) \left(1 - \rho_{im+1}^k \rho_{im+2}^k \cdots \rho_{il}^k\right)$$

(3.14)

Algorithm 1. Calculate α_{im}^k in CFS_i^k

1: $\beta_{im}^k \leftarrow 0$, $temp \leftarrow 0$, $A \leftarrow \varnothing$
2: **for** all node m in $CFS_i{}^k$ **do**
3: Calculate the probability $\beta_{im}^k(1)$ according to (3.8)
4: Calculate the probability $\beta_{im}^k(2)$ according to (3.9)
5: $\beta_{im}^k \leftarrow \beta_{im}^k(1) + \beta_{im}^k(2)$
6: $temp \leftarrow \beta_{im}^k + temp$
7: **end for**
8: **for** all node m in $CFS_i{}^k$ **do**
9: $\alpha_{im}^k \leftarrow \beta_{im}^k / temp$
10: $A \leftarrow A \cup \{\alpha_{im}^k\}$
11: **end for**
12: **return** A

Note that our scheme adopts network coding (NC) and the coding operations are similar to the intra-session NC in MORE. Using intra-session NC, intermediate nodes perform NC operation on packets of the same session.

And, similar to the WACA, we maintain $|K|$ credit counters for each node. Each credit counter corresponds to a channel. In our scheme, we consider the impact of channel availability of CRNs on credit calculating, which is shown as

$$\text{credit}_i^k = \frac{P_i^k}{\sum\limits_t \left(\sum\limits_{<g,i> \in \psi_i^{k+}} f_{gi}^k \times \left(1 - \rho_{gi}^k \right) \times \mu_{gi}^k(t) \right)}, i, g \in V, i \neq S \qquad (3.15)$$

In Eq. (3.15), the parameter $\mu_{gi}^k(t)$ is the channel availability depending on PU's behavior in CRNs. If the credit_i^k becomes positive, the node creates a coded packet, broadcasts it on channel k, and then decrements the credit counter.

In this paper, our goal is to maximize the aggregate throughput at destination node D, which is a nonlinear programming (NLP) problem. Putting all the above constraints together, the objective function of the formulation is expressed as

$$\max \sum_k \sum_{<j,D> \in \psi_D^{k+}} f_{jD}^k \times \left(1 - \rho_{jD}^k \right) \qquad (3.16)$$

s.t.

$$
\begin{cases}
0 \leq u_{ij}^k(t) \leq 1, <i,j> \in E \\
\sum\limits_{<m,n> \in I_{ij}} \mu_{mn}^k(t) \leq 1, <m,n> \in I_{ij}, <i,j> \in E \\
u_{ij}^k(t) \leq \theta_i^k(t), <i,j> \in \psi_i^{k-}, i,j \in V, i \neq D \\
u_{gi}^k(t) \leq \theta_i^k(t), <g,i> \in \psi_i^{k+}, g,i \in V, i \neq S \\
\sum\limits_k \theta_i^k(t) \leq R \\
0 \leq \theta_i^k(t) \leq 1 \\
0 \leq \alpha_{gi}^k \leq 1 \\
0 < \rho_{ij}^k < 1 \\
u_{ij}^k(t) \leq h_{ij}^k, <i,j> \in E \\
h_{ij}^k \in \{0,1\} \\
0 \leq f_{ij}^k \leq \sum\limits_t u_{ij}^k(t) \times B, <i,j> \in E \\
\sum\limits_k \sum\limits_{<g,i> \in \psi_i^{k+}} f_{gi}^k \times \left(1 - \rho_{gi}^k \right) \times \alpha_{gi}^k = \sum\limits_k P_i^k + Q_i(T), g,i \in V, i \neq S \\
\forall k \in K, t \in T, i,j,g \in V
\end{cases} \qquad (3.17)
$$

3.3.1.3 Heuristic Algorithm

Even though the LP can be used to obtain optimal forwarding candidates and channel assignment, it takes a substantial amount of time to terminate in large-scale CRNs. Hence, we present a heuristic algorithm to select forwarding candidates and assign channel in CRNs.

The idea behind the heuristic algorithm is to select a set of nodes and channels for each time slot such that the throughput is maximum, which includes candidate selection algorithm and channel assignment algorithm.

Candidate Selection and Prioritization

In CRNs, after detecting an idle channel, the sender needs to select a candidate for data dissemination.

Considering the queue backlog of node i, we use the summation of queue backlog and ETX as the forwarding candidate selector criterion, which can be expressed as

$$H_i(T) = \chi Q_i(T) + \gamma \text{ETX}_i \tag{3.18}$$

where χ and γ are the weights, constrained by $\chi + \gamma = 1$, which are set to be 0.5 and 0.5 in our simulations. We further explore the trade-offs offered by χ in Section 6.

In Eq. (3.18), we consider the queue length and channel availability in forwarding scheme. The smaller $H_i(T)$ node i has, the higher is the probability of the node to be selected as a forwarding candidate. If the queue of a node is almost full, which means packet loss will occur at the node, the node should not receive more packets.

Selection of the forwarding nodes must obey the following conditions: The $H_i(T)$ between the sender and candidates must be below threshold $H_{\text{threshold}}(T)$ limit to ensure good quality link. In practice, the threshold $H_{\text{threshold}}(T)$ is given by

$$H_{\text{threshold}}(T) = \frac{1}{|\text{Neighbor}|} \sum_{i \in \text{Neighbor}} H_i(T) \tag{3.19}$$

where Neighbor is the set of node i's one-hop neighboring nodes.

The **preconditions** that a node can be selected into a CFS are as follows:

1. The node and the sender (the node's previous-hop node) share at least one channel; also, the node and its next-hop node share at least one channel.
2. The $H_i(T)$ value of the node is lower than the threshold $H_{\text{threshold}}(T)$.
3. It is a direct neighboring node of the sender.

Algorithm 2. Candidate Selection Algorithm

1: $CFS_v \leftarrow \emptyset$
2: **for** all node $j \in Neighbor(v)$ **do**
3: Calculate its $H_j(T)$ and the threshold $H_{threshold}(T)$ according to (3.13), (3.14)
4: **if** $\left(H_j(T) \leq H_{threshold}(T)\right)$ && $(Ch(v) \cap Ch(j) \neq \emptyset)$ **then**
         && $(Ch(j) \cap Ch(Neighbor(j)) \neq \emptyset)$
5: $CFS_v \leftarrow CFS_v \cup j$
6: **end if**
7: **end for**
8: **return** CFS_v

The candidate selection algorithm for any node v (except for destination node) is listed in **Algorithm 2**, where Neighbor (v) is the set of v's next-hop nodes, Ch (v) is the available channel set of node v, and CFS_v is the forwarding candidate set of node v.

When a sender is ready to send a packet, it inserts an extra header into this packet, which lists all nodes in CFS. Nodes in the *CFS* are ranked according to their $H_i(T)$. The one with smaller $H_i(T)$ value to the destination has higher priority.

Channel Assignment

After candidate selection over multichannel CRNs, we must choose and allocate the channel from the available channel set to the radio of a node. Channel assignment can either be centralized [41, 42] or distributed [43, 44] over CRNs. A centralized approach to the channel assignment usually obtains best results. However, it typically brings about a high communication overhead. In addition, the channel availability of CRNs is frequently time-varying; the centralized policies become less efficient. Thus, we design a distributed approach in our solution. Our scheme jointly considers the transmission time and available time of an available channel, which is shown in **Algorithm 3**.

According to our work [45], we can obtain the transmission time T_r and the available time T_v of a given channel. Each node will assign the first channel (that has minimal Δ) to its first radio and the second channel to its second radio and so on. We assume that if a channel is assigned, it is immediately used to transmit data packets. Note that if a PU arrives on the assigned channel, then only the affected nodes will update the available channel set and repeat the **Algorithm 3**.

Algorithm 3. Channel Assignment Algorithm

---

i: the node in CRNs which runs **Algorithm 3**

Neighbor: set of i's neighbors

$Ch(i)$: set of node i's available channels

ChN: Set of i-neighboring nodes' available channels

R: Set of radios on node i

X: Set of channels in $Ch(i)$ which are not assigned

Z: Set of radios in R on which channel is not assigned

$C(R)$: Set of channels tuned to set of radios R

T_r :Transmission time of a given channel

T_v: Available time of a given channel

1: **Input**: i, *Neighbor*, $Ch(i)$, ChN, R

2: $Z \leftarrow R$, $X \leftarrow Ch(i)$, $Temp \leftarrow Ch(i) \cap ChN$

3: **while** *Temp* $\neq \varnothing$ **do**

4: Sort *Temp* by value of $\Delta = T_v - T_r$ in ascending order

5: Update *Temp* with removing the channels that have negative value of Δ

6: **while** $Z \neq \varnothing$ **do**

7: Select a node from i' *Neighbor*

8: Select a channel c that has minimal Δ

9: $C(R) \leftarrow c, Z \leftarrow Z/\{z\}, X \leftarrow X/\{c\}$

10: **end while**

11: **end while**

12: **Output** $C(R)$

3.3.1.4 Simulation and Performance Evaluation

In this section, we evaluate the performance of CANCOR by simulation under different network settings, e.g., the number of radios/channels, buffer size, and batch size, using NS2 [46] and CRCN module [47]. We set up a CRN with 4 PUs and 30 SUs randomly distributed in a 1000×1000 m^2 area. The transmission range, d_s, and corresponding interference range, d_I, of each SU are set to be 250 and 500 m. The PU coverage is 550 m. The channel data rate is 2 Mbps and CBR rate is 800Kbps. The channel sensing time is 5 ms and the channel changing time is 80 μs. The packet size is 1000 bytes. We set the packet loss rate $\rho_{ij}^k \in [0.1, 0.3]$ and the mean rate $\mu_p \in [0.2, 0.45]$. Our simulations are averaged over 100 times, each lasts for 350 s. The batch size is ten packets. The PU on time is 5 s. The buffer size is 100 KB. We compare the proposed scheme, CANCOR, with the following protocols: shortest path routing (SINGLE), ExOR, MORE, MaxPoS in terms of throughput.

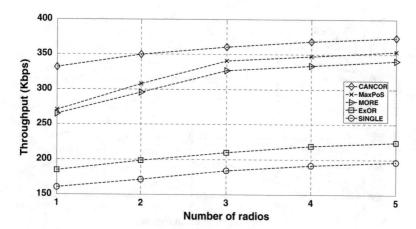

Fig. 3.3 Throughput comparison vs. number of radios

In the following, we study performance impact of the number of radios/channels, buffer size, batch size, and PU activity on throughput and PU harmful interference ratio in CANCOR and other routing schemes.

The Performance Impact of Multi-radio

The throughput of each routing scheme under different radios is shown in Fig. 3.3. We fix the number of channels to be 5. As can be seen from Fig. 3.3, all the routing protocols are able to exploit the increase in the number of radios to obtain a solution with improved throughput. This is because using more radios allows more simultaneous transmissions in CRNs. The ability to exploit more simultaneous transmissions on different nonoverlapping channels reduces the interference and improves the network throughput. In addition, we see that the throughput increases significantly from 1 radio to 2 radios and from 2 to 3 radio cases, much more than the percentage increase from 3 to 4 and from 4 to 5 radio cases. In addition, the CANCOR is better than the other routing schemes, which exploits the novel routing metric and candidate selection algorithm.

The Performance Impact of Multichannel

In this evaluation, we varied the number of channels in CRNs to study its impact on the throughput. We fixed the number of radios to be 3. As expected, we observe the trend that as the number of channels increase, the throughput generally increases, as shown in Fig. 3.4. This is because as the number of channels increases, there would be more available channels for SU transmission, so the throughput of each scheme increases. CANCOR, MaxPoS, ExOR, and MORE achieve higher throughput than

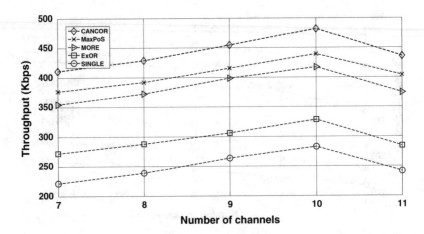

Fig. 3.4 Throughput comparison vs. number of channels

SINGLE. The reason is that these four schemes take advantage of the inherent property of OR-opportunistic forwarding by using multiple forwarding candidates, while SINGLE always uses the same route consisting of a forward candidate. CANCOR performs better than ExOR, MaxPoS, and MORE. This is because we exploit a new method for selecting forwarding candidates for CANCOR in CRNs, which considers the queue length and channel availability. Also, network coding in CANCOR can reduce the retransmissions over forwarders' data transmission. In addition, we remark that the throughput may not always increase when the number of channels increases. This is because the proposed channel assignment algorithm is not necessarily optimal. As the number of channels increases, the probability that two interfering links share the same channels increases as well, because our scheme does not consider interference in channel assignment. Moreover, when the number of channels increases, the probability that one node belongs to multiple CFS (candidate forwarders set) will be larger, so the buffer of the node will overflow, inducing a lot of packet loss. As a result, the throughput will decrease when the number of channels increases to a certain value. In our simulation, when the number of channels increases up to 10, the throughput will be maximum and then the throughput decreases with the increase of the number of channels. The reason is that a lot of packets drop due to some nodes belonging to multiple CFS overflow and channel interference when the number of channels reaches 10.

The Performance Impact of Buffer Size

In Fig. 3.5 ($R = 3$, $K = 4$), we can see that as buffer size increases, the throughput increases, and finally the growth is slowed down. This is due to the fact that for large buffer size, the packet loss is low. However, when the buffer size goes up to a certain value, the throughput increases slowly, as shown in Fig. 3.5. It is observed

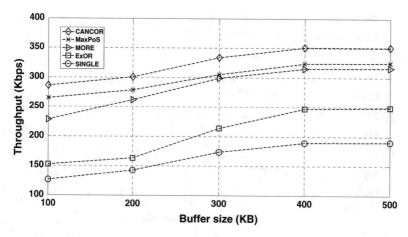

Fig. 3.5 Throughput comparison vs. buffer size

that the CANCOR achieves much higher throughput than MORE, MaxPoS, ExOR, and SINGLE. This is because it considers the queue length and channel availability for selecting forwarding candidates in CANCOR; also it exploits network coding technology in CANCOR, which can reduce the retransmissions over multi-hop CRNs.

The Performance Impact of Batch Size

In this evaluation, the number of radios is 3 and the number of channels is 4. We explore the performance of CANCOR, MORE, and ExOR for various batch sizes. In Fig. 3.6, we can see that as batch size is increasing, the throughput of the three schemes increases. However, the network coding-based schemes (CANCOR and MORE) are insensitive to batch size. The ExOR can receive a significant throughput gain when increasing batch size.

3.3.2 CROR: Coding-Aware Opportunistic Routing in Cognitive Radio Networks

Cognitive radio is a promising technology to improve spectrum utilization. However, spectrum availability is uncertain which mainly depends on primary user's (PU's) behaviors. This makes it more difficult for most existing CR routing protocols to achieve high throughput in multichannel CRNs. Intersession network coding and opportunistic routing can leverage the broadcast nature of the wireless channel to improve the performance for CRNs. In this section, we present a coding-aware

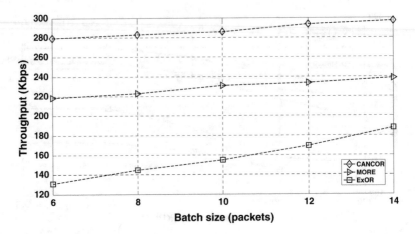

Fig. 3.6 Throughput comparison vs. batch size

opportunistic routing protocol for multichannel CRNs, cognitive radio opportunistic routing (CROR) protocol, which jointly considers the probability of successful spectrum utilization, packet loss rate, and coding opportunities. We evaluate and compare the proposed scheme against three other opportunistic routing protocols with multichannel. It is shown that the CROR, by integrating opportunistic routing with network coding, can obtain much better results, with respect to throughput, the probability of PU-SU packet collision, and spectrum utilization efficiency.

Network coding (NC) is another technique to increase the throughput of both wired and wireless networks. However, none of the existing works systematically combines NC and OR to improve the performance of multichannel CRNs. Figure 3.7 shows an example of network coding-aware opportunistic routing. In Fig. 3.7, node 0 wants to send packet P1 to node 2 through path $0 \rightarrow 1 \rightarrow 2$, and node 7 wants to send packet P2 to node 5 through path $7 \rightarrow 6 \rightarrow 5$. These two data transmissions are denoted by green arrow lines. When node 0 sends P1, nodes 1, 3, 4, and 5 will receive/overhear P1 denoted by black arrow lines. Similarly, when node 7 sends P2, nodes 2, 3, 4, and 6 will receive/overhear P2 denoted by black arrow lines. In traditional routing (e.g., AODV, DSR), except for nodes 1 and 6, the remaining nodes will drop the packet received/overheard, while in coding-aware OR, it allows nodes 3 and 4 to store packets received/overheard and employ the opportunistic coding approach [8] to code together. Each node in the forwarding set forwards any packets that have still not been received by any higher priority node. If the priority of node 3 is the highest, node 3 XORs P1 and P2, and broadcasts the coded packet P1 ⊕ P2, nodes 5 and 2 are able to obtain their needed packets. It requires one additional transmission for transmitting two native packets to nodes 2 and 5 without NC. In other words, NC can reduce the number of transmissions in multi-hop wireless networks. Several protocols based on NC scheme have been proposed. COPE [8] is the first practical network coding mechanism for supporting efficient unicast communication in wireless mesh networks (WMNs). In COPE, each node overhears the communications that take place in its neighborhood and records the packets that have been received by the neighbor nodes. In CORE [15], packets of multiple flows are coded together and are transmitted to the candidate next hops. The priority of each candidate is determined in a distributed way based on a utility metric. COPE and CORE exploit intersession network coding to improve the throughput performance of wireless mesh networks.

All abovementioned coding-aware routing protocols are based on single-channel networks. It may not be suitable for multichannel networks, specifically multi-hop CRNs, where the channel availability is uncertain. Note that in our scheme, we mainly consider joint design of intersession NC and OR in CRNs. Joint design of intra-session NC and multichannel is also an interesting area. However, it is technically irrelevant to our topic.

3.3.2.1 System Model

We assume an interweave model, i.e., the nodes in the CRNs can only transmit when the PUs are not active. In this paper, we consider a time-slotted multi-hop cognitive radio network (CRN) with M ($M \geq 2$) licensed orthogonal channels. There are *nums* SUs and *nump* PUs in this CRN. We assume every node (including SU and PU) is equipped with two radios, one for data transmission and the other for control signals. Each SU is capable of sensing the locally available channels and has the capability of channel changing at packet level for data transmission.

In CRNs, the SUs can opportunistically exploit the channels when the PUs currently do not occupy. When PU appears on a channel, the SU should vacate it for the PU. The SU's communication mainly depends on the PU's activities, which is the biggest difference between CRNs and traditional wireless networks, e.g., ad hoc networks and mesh networks. The usage pattern of PUs over a given channel m that is available for SU transmissions follows an independent ON/OFF state model [38]. The ON period T_{on} (m) represents the time that the PUs are occupying channel m, and the OFF period T_{off} (m) represents the time that the PUs are inactive on channel m.

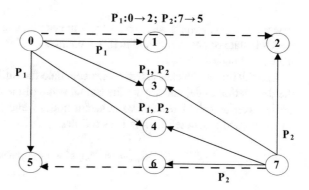

Fig. 3.7 An example of coding-aware OR

3.3.2.2   CROR: Coding-Aware Routing in CRNs

The basic idea behind CROR is as follows: first, we propose a new routing metric
SuDR, considering available time and packet loss rate of a given channel, for
forwarding candidate selection. At each node of a forwarding candidate set (FCS),
the packets going to different next hops are able to decode the resulted packet, and
they all share the same channel. The CROR protocol includes the following four
major parts: (1) new routing metric SuDR, (2) candidate selection and prioritization,
(3) coding rules and modeling assumptions, and (4) set forwarding timer.

A New Routing Metric: SuDR

We propose a new metric called successful delivery ratio (SuDR), which captures
the number of transmissions needed to deliver a packet for multi-hop CRNs.

MORE and CORE use expected transmission count (ETX) to select forwarding
candidate; however, in CRNs, due to PU activities, the routing metric in CRNs must
be aware of channel availability at intermediate node. Next, we give the method to
calculate SuDR.

The required transmission time for a data packet of size L in channel m between
any two neighboring node i and node j (SUs) can be calculated as follows:

$$T_{trij}(m) = L/R_{ij}(m) \tag{3.20}$$

where $R_{ij}(m)$ is the transmission rate of channel m.

According to reference [48], we can obtain the channel available time $T_{av\_ij}(m)$,
which is exponentially distributed. Thus, the probability $p_{ij}(m)$ that the channel m
will be selected successfully for transmission between node i and node j can be
represented as

$$p_{ij}(m) = P\left[T_{avij}(m) \geq T_{trij}(m)\right]$$
$$= 1 - F_{T_{avij}(m)}\left(T_{trij}(m)\right) = e^{-\frac{T_{trij}(m)}{u_{ij}(m)}} \tag{3.21}$$

where $F_{T_{avij}(m)}\left(T_{trij}(m)\right) = 1 - e^{-\frac{T_{trij}(m)}{u_{ij}(m)}}$ is the cumulative density function (CDF) of
the OFF state of channel m and $u_{ij}(m)$ is channel availability of channel m between
node i and node j.

Considering the packet loss rate, we calculate the SuDR of a node starting from
the destination node, and we set the SuDR value of the destination node as 1. The
$SuDR_{ij}$ can be calculated in two adjacent nodes i and j, i.e., the number of hops
between i and j, denoted as h, is 1, as follows:

$$SuDR_{ij}(m) = p_{ij}(m) \times \left(1 - \rho_{ij}(m)\right) \tag{3.22}$$

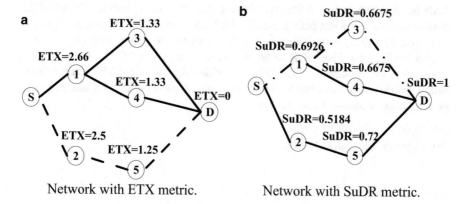

Fig. 3.8 Network with different routing metrics

$$\text{SuDR}_{ij}[h] = \max_{m \in M}\left\{\text{SuDR}_{ij}(m)\right\} \quad (h = 1) \tag{3.23}$$

where M is the set of common available channels for node i and node j and $\rho_{ij}(m)$ is the packet loss rate on link between node i and node j on channel m.

When the number of hops between i and j, h is greater than or equal to 2, the SuDRij can be recursively expressed as

$$\text{SuDR}_{ij}[h] = 1 - \prod_{r=1}^{N}\left(1 - \text{SuDR}_{ir}\text{SuDR}_{rj}[h-1]\right) \quad (h \geq 2) \tag{3.24}$$

where r is the node i's neighbor and N is the number of node i's neighbors. The number of hops between r and j is $(h-1)$. Note that the packet forwarding is independent to previous forwarding.

In the following, we give an example to illustrate how the candidate can be selected with ETX and SuDR, respectively. Assume that there are seven nodes (SUs) and three licensed orthogonal channels in CRNs; the network topology is shown in Fig. 3.8. The available channel set is M, $M = \{\text{Ch1}, \text{Ch2}, \text{Ch3}\}$. The available channels of each node are S = [Ch2, Ch3], 1 = [Ch2, Ch3], 2 = [Ch1, Ch3], 3 = [Ch2, Ch3], 4 = [Ch2, Ch3], 5 = [Ch1, Ch3], and D = [Ch1, Ch2, Ch3]. For simplicity, we assume that the probability $p_{ij}(m)$ is equal between any two neighboring nodes over channel m, which is denoted as $p(m)$, and also assume that $p(m) = \{p(1), p(2), p(3)\} = \{0.9, 0.92, 0.89\}$. Similarly, $\rho_{ij}(m)$ can be expressed as $\rho(m)$ over channel m, assuming that $\rho(m) = \{\rho(1), \rho(2), \rho(3)\} = \{0.2, 0.3, 0.25\}$.

According to ETX metric and the forwarding scheme, we can obtain the final value of ETX of each node (except for source node S, as shown in Fig. 3.8a). Similarly, we can obtain the final value of SuDR of each node (except for source node S, as shown in Fig. 3.4b). We observe that the routing metric will affect the path selection. We select path S-2-5-D with ETX, denoted by red line as shown in Fig. 3.4a. However, we select path S-1-3-D with SuDR, denoted by green line as shown in Fig. 3.8b. The SuDR value of a node is its successful delivery ratio from the node to the destination considering the channel availability. The scheme using SuDR metric can achieve better performance than the ones with ETX metric as will be seen in our simulations results later:

(a) Network with ETX metric
(b) Network with SuDR metric

Candidate Selection and Prioritization

After detecting an idle channel, the sender needs to select a candidate for data dissemination. Selection of the forwarding nodes must obey the following conditions. The SuDR between the sender and candidates must be below threshold SuDRc limit to ensure of good quality link.

The preconditions that a node can be selected into a FCS are as follows:

1. The node and the sender (the node's previous-hop node) share at least one channel; also, the node and its next-hop node share at least one channel.
2. The SuDR value of the node is higher than the threshold.
3. It is a direct neighboring node of the sender.
4. The nodes which are in the FCS must be able to mutually overhear each other with a certain probability.

The candidate selection algorithm for any node v (except for destination node) is listed in **Algorithm 4**, where $N(v)$ is the set of v's next-hop nodes, Ch (v) is the available channel set of node v, and FCSv is the forwarding candidate set of node v.

Algorithm 4. Candidate Selection Algorithm

1: function Selection(v, FCSv, SuDRc)
2: $FCS_v \leftarrow \varnothing$
3: for all node $j \in N(v)$ do
4: Calculate its SuDR according (3.17), (3.18), (3.19)
5 if $\left(SuDR_j \geq SuDR_c\right)$ && $(Ch(v) \cap Ch(j) \neq \varnothing)$ && $(Ch(j) \cap Ch(N(j)) \neq \varnothing)$ then
6: $FCS_v \leftarrow FCS_v \cup j$
7: end if
8: end for
9: return FCSv

When a sender is ready to send a packet, it inserts an extra header into this packet, which lists all nodes in FCS. Nodes in the FCS are ranked according to their SuDR. The one with larger SuDR value to the destination has higher priority. Thus, the node will forward packets earlier, and other nodes hearing this forwarding will cancel their timers.

Coding Rules and Modeling Assumptions

Assume packet P_i is at node v, $u(P_i)$ is the set of the previous-hop node of packet P_i, $n(P_i)$ is the set of the next-hop node of packet P_i, Sv (P_i) is the state of packet P_i at node v, and Ch (v) is the available channel set of node v. If Sv $(P_i) = n$, it means that node v receives the packet P_i in normal fashion. If Sv $(P_i) = d$, it means node v has decoded packet P_i. Consider k packets $P_1, P_2, \ldots, P_k$ at node v that have distinct next-hop nodes $n_1, n_2, \ldots, n_k$, respectively. Based on [49], we modify the coding rules, which consider the channel availability and the number of encoding packets. Suppose these are coded together to form the coded packet $P = P_1 \oplus P_2 \oplus \ldots \oplus P_k$ that is broadcast to all the above next-hop nodes. This is a valid network coding if the next-hop node n_i for each packet P_i already has all other packets P_j for $j \neq i$ (so that it can decode P_i)—this can occur if

$$n(P_i) = u(P_j) \tag{3.25}$$

$$\left(n(P_i) \in N(u(P_j))\right) \cap S_v(P_j) = n \tag{3.26}$$

$$\text{Ch}(n(P_i)) \cap \text{Ch}(n(P_j)) \neq \varnothing \tag{3.27}$$

In CROR, all nodes are set in the promiscuous mode; they can overhear packets not addressed to them. Both encoding and decoding are XOR operations. The encoding algorithm is listed in **Algorithm 5**. The parameter average encoding number, AvgCodingNos, is used to determine how many packets we can encode for practical wireless network coding, XOR, which we can obtain from reference [50].

Algorithm 5. Encoding Algorithm

1: Capable ← True; AvgCodingNo ← 0; AvgCodingNos ← 7
2: CodedPacket ← front packet of Qi
3: while Capable do
4: Capable ← False
5: for all packet Pj in Qi do
6: Check the coding conditions according (3.20), (3.21), (3.22)
7: If it satisfies all coding conditions then
8: Capable ← True

9: If forwarding timer is not expired && Pj is not in CodedPacket then
10: CodedPacket ← Pj ⊕ CodedPacket
11: AvgCodingNo++
12: end if
13: end if
14: If AvgCodingNo ≥ AvgCodingNos then
15: Capable ← False; Break
16: end if
17: end for
18: Mark latest include native packet as encoded
19: end while
20: return CodedPacket

Set Forwarding Timer

Forwarding timer is the most important aspect of CROR, which is used to avoid
redundant transmission, but it also affects the overall throughput of the CROR.
After receiving a packet, every node sets a timer for each packet. The value of the
timer is proportional to the reciprocal of the product of the number of packets that
can be coded at the node and the SuDR value. Thus, the node with more coding
opportunities and higher SuDR value will forward the packet earlier, and the other
nodes hearing this forwarding will cancel their timers.

Let y be the number of packets that can be coded at the node j, and l is the order
of the node in the FCS. Then the forwarding timer t will be

$$t = \frac{l}{SuDR_j \times y^2} \tag{3.28}$$

3.3.2.3 Performance Evaluation

In this section, we evaluate the performance of CROR protocol by simulation
under different network settings, e.g., PU activity, packet loss rate, and bandwidth
difference in spectrum switch state, using NS2 [46] and CRCN model [47]. We set
up a CRN with 9 PUs and 32 SUs randomly distributed in a 1500×1500 m^2 area.
As mentioned earlier, the PU activity in each channel is modeled as an exponential
ON-OFF process; thus, the channel availability of channel m, $\mu_{ij}(m)$ is selected
accordingly. The network parameter settings are shown in Table 3.2. We compare
the following three protocols, ExOR, MORE, and MaxPoS, in terms of throughput,
the probability of PU-SU packet collision, and bandwidth efficiency.

Table 3.2 Simulation parameters

Number of channels	3
$\mu_{ij}(m)$ $(m = 1, 2, 3)$	{0.3; 0.5; 0.7}
Number of PUs per channel	3
Number of SUs	32
PU coverage	550 m
SU transmission range	250 m
Channel data rate	2 Mbps
CBR rate	800 kbps
Per channel sensing time	5 ms
Channel changing time	70 μs
Packet size	1000 bytes
E[Toff]	[100, 700 ms]
AvgCodingNos	7
SuDR threshold	0.55

Impact of Packet Loss Rate on Throughput

Next, we evaluate the impact of packet loss rate on throughput. In this test, we vary the packet loss rate between any two nodes which are neighbors per channel. Figure 3.9 shows that as the packet loss rate increases, the throughput decreases in all routing protocols. This is because the larger the packet loss rate, the more packets should be retransmitted. However, the CROR still outperforms other three schemes with higher throughput, which simultaneously exploits channel availability and network coding in routing design. In addition, we can observe that when packet loss rate is larger, the opportunistic routing protocol (CROR, MORE)-based network coding has better performance than that of the protocols without network coding, which means that network coding still brings coding gain in higher packet loss rate with fewer number of transmissions.

Impact of Spectrum Changing on Bandwidth Utilization Efficiency

The impact of spectrum changing on bandwidth utilization efficiency is depicted in Fig. 3.10. On PUs arrival, the affected SUs should switch channel for continuously transmitting data packets. In this process, the bandwidth may change, which mainly affects the transmission time and packet loss rate. We investigate the bandwidth efficiency of the four schemes in this scenario: three channels, Ch1, Ch2, and Ch3, having varying raw channel bandwidth, 1, 2, and 4 Mbps. The switch sequences are 2 Mbps → 1 Mbps, 4 Mbps → 2 Mbps, and 4 Mbps → 1 Mbps. The bandwidth differences are 1, 2, and 3 Mbps. The number of PU appearance is 1 and the PU on time is 1 s. As can be seen in Fig. 3.8, as the bandwidth difference grows, the bandwidth utilization efficiency drops. However, CROR has higher bandwidth

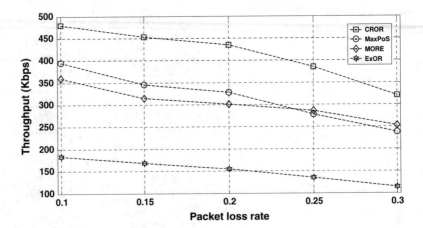

Fig. 3.9 Throughput vs. packet loss rate

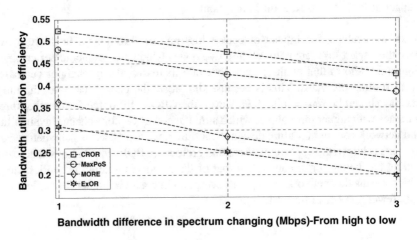

Fig. 3.10 Bandwidth efficiency vs. difference in spectrum changing

efficiency than other three schemes, implying that our scheme is effective in fully utilizing the spectrum resource, which considers the channel available time, transmission time, and coding opportunities in routing design.

3.4 Conclusions

In this chapter, we first survey some classical routing protocols that incorporated network coding with routing schemes. At the same time, their representative features and drawbacks are analyzed. We also introduce several multicast schemes with network coding at network layer. In addition, two routing protocols that exploit network coding technology to improve routing performance for CRNs are proposed. The simulation results show that the proposed schemes can achieve better performance.

Acknowledgment This work was supported by the Science and Technology Fundament Research Fund of Shenzhen under grant JCYJ20140417172417131 and the Natural Science Foundation of Guangdong Province under grant 2014A030313698.

References

1. R.W. Yeung, Z. Zhang, Distributed source coding for satellite communications. IEEE Trans Inf Theory **45**(4), 1111–1120 (1999)
2. R. Ahlswede, N. Cai, S.-Y.R. Li, R.W. Yeung, Network information flow. IEEE Trans Inf Theory **46**(4), 1204–1216 (2000)
3. S. Li, R. Yeung, N. Cai, Linear network coding. IEEE Trans Inf Theory **49**(2), 371–381 (2003)
4. R. Koetter, M. Medard, An algebraic approach to network coding. IEEE/ACM Trans Networking **11**(5), 782–795 (2003)
5. T. Ho, M. Medard, R. Koetter, D. Karger, M. Effros, J. Shi, B. Leong, A random linear network coding approach to multicast. IEEE Trans Inf Theory **52**(10), 4413–4430 (2006)
6. A. Ramamoorthy, J. Shi, R. Wesel, On the capacity of network coding for random networks. IEEE Trans Inf Theory **51**(8), 2878–2885 (2005)
7. D. Lun, N. Ratnakar, R. Koetter, M. Medard, E. Ahmed, H. Lee, Achieving minimum-cost multicast: a decentralized approach based on network coding, in *Proceedings of the IEEE INFOCOM*, vol. 3, 2005, pp. 1607–1617
8. S. Katti, H. Rahul, W. Hu, D. Katabi, M. Medard, J. Crowcroft, XORs in the air: practical wireless network coding. IEEE/ACM Trans Networking **16**(3), 497–510 (2008)
9. S. Chachulski, M. Jennings, S. Katti, D. Katabi, Trading structure for randomness in wireless opportunistic routing, in *Proceedings of the ACM SIGCOMM*, 2007, pp. 169–180
10. J. Le, J.C.S. Lui, D.-M. Chiu, DCAR: distributed coding-aware routing in wireless networks. IEEE Trans Mob Comput **9**(4), 596–608 (2010)
11. T. Ho, D.S. Lun, *Network Coding: An Introduction* (Cambridge University Press, New York, 2008)
12. H. Seferoglu, A. Markopoulou, K.K. Ramakrishnan, I2NC: Intro- and inter-session network coding for unicast flows in wireless networks, in *Proceedings of the IEEE Infocom*, 2011
13. B. Guo, H. Li, C. Zhou, Y. Cheng, Analysis of general network coding conditions and design of a free-ride-oriented routing metric. IEEE Trans Veh Technol **60**(4), 1714–1727 (2011)
14. S. Biswas, R. Morris, ExOR: opportunistic multi-hop routing for wireless networks, in *Proceedings of the ACM SIGCOMM*, vol. 35, 2005, pp. 133–144
15. Y. Yan, B. Zhang, J. Zheng, J. Ma, CORE: a coding-aware opportunistic routing mechanism for wireless mesh networks. IEEE Wirel Commun **17**(3), 96–103 (2010)
16. M. Aajami, H. Park, J. Suk, Combining opportunistic routing and network coding: a multi rate approach, in *Proceedings of the IEEE WCNC*, 2013, pp. 2208–2213
17. D. Koutsonikolas, Y. Hu, C. Wang, XCOR: synergistic interflow network coding and opportunistic routing, in *Proceedings of the ACM MobiCom*, 2008, pp. 1–3
18. Q. Hu, J. Zheng, CoAOR: an efficient network coding aware opportunistic routing mechanism for wireless mesh networks, in *Proceedings of the IEEE GLOBECOM*, 2013, pp. 4578–4583
19. Y. Lin, B. Li, B. Liang, CodeOR: opportunistic routing in wireless mesh networks with segmented network coding, in *Proceedings of the IEEE ICNP*, 2008, pp. 1092–1648

20. D. Koutsonikolas, C. Wang, Y. Hu, CCACK: efficient network coding based opportunistic routing through cumulative coded acknowledgments, in *Proceedings of the IEEE INFOCOM*, 2010, pp. 216–224
21. J. Jin, H. Xu, B. Li, Multicast scheduling with cooperation and NetworkCoding in cognitive radio networks, in *Proceedings of the INFOCOM*, 2010, pp. 1289–1297
22. L. Chen, T. Ho, S.H. Low, M. Chiang, J.C. Doyle, Congestion control for multicast flows with network coding. IEEE Trans Inf Theory **58**(9), 5908–5921 (2012)
23. H. Wu, J. Zheng, CoRET: a network coding based multicast retransmission scheme for mobile communication networks, in *IEEE International Conference on ICC*, 2011, pp. 1–5
24. H.-C. Lu, W. Liao, Cooperative strategies in wireless relay networks. IEEE J Select Areas Commun **30**(2), 323–330 (2012)
25. P. Li, S. Guo, S. Yu, A.V. Vasilakos, Reliable multicast with pipelined network coding using opportunistic feeding and routing. IEEE Trans Parallel Distrib Syst **25**(12), 20–24 (2014)
26. S. Haykin, Cognitive radio: brain-empowered wireless communications. IEEE J Select Areas Commun **23**(2), 201–220 (2005)
27. M. Pan, R. Huang, Y. Fang, Cost design for opportunistic multi-hop routing in cognitive radio networks, in *Proceedings of the IEEE MILCOM*, 2008, pp. 1–7
28. H. Khalife, S. Ahuja, N. Malouch, M. Krunz, Probabilistic path selection in opportunistic cognitive radio networks, in *Proceedings of the IEEE GLOBECOM*, 2008, pp. 1–5
29. O.S. Badarneh, H.B. Salameh, Opportunistic routing in cognitive radio networks: exploiting spectrum availability and rich channel diversity, in *Proceedings of the IEEE GLOBECOM*, 2011, pp. 1–5
30. S. -C. Lin, K.-C. Chen, Spectrum aware opportunistic routing in cognitive radio networks, in *Proceedings of the IEEE GLOBECOM*, 2010, pp. 1–6
31. Y. Liu, L.X. Cai, X. Shen, J. W. Mark, Exploiting heterogeneity wireless channels for opportunistic routing in dynamic spectrum access networks, in *Proceedings of the IEEE ICC*, 2011, pp. 1–5
32. Y. Liu, L.X. Cai, X. Shen, Spectrum-aware opportunistic routing in multi-hop cognitive radio networks. IEEE J Select Areas Commun **30**(10), 1958–1968 (2012)
33. F. Wu, N. Vaidya, Workload-aware opportunistic routing in multi-channel, multi-radio wireless mesh networks, in *Proceedings of the IEEE SECON*, 2012, pp. 344–352
34. S. He, D. Zhang, K. Xie, H. Qiao, J. Zhang, A simple channel assignment for opportunistic routing in multi-radio multi-channel wireless mesh networks, in *Proceedings of the IEEE MSN*, 2011, pp. 201–208
35. M.H. Rehmani, A.C. Viana, H. Khalife, S. Fdida, SURF: a distributed channel selection strategy for data dissemination in multi-hop cognitive radio networks. Elsevier Comput Commun **36**(10), 1172–1185 (2013)
36. B. Mumey, J. Tang, I.R. Judson, D. Stevens, On routing and channel selection in cognitive radio mesh networks. IEEE Trans Veh Technol **61**(9), 4118–4128 (2012)
37. A. Goldsmith, S.A. Jafar, I. Maric, S. Srinivasa, Breaking spectrum gridlock with cognitive radios: an information theoretic perspective. Proc IEEE **97**(5), 894–914 (2009)
38. H. Kim, K.G. Shin, Efficient discovery of spectrum opportunities with MAC-layer sensing in cognitive radio networks. IEEE Trans Mob Comput **7**(5), 533–545 (2008)
39. D. S. J. De Couto, D. Aguayo, J. Bicket, R. Morris, A high-throughput path metric for multi-hop wireless routing, in *Proceedings of the ACM MOBICOM*, 2003, pp. 134–146
40. P. Gupta, P.R. Kumar, The capacity of wireless networks. IEEE Trans Inf Theory **46**(2), 388–404 (2000)
41. V. Brik, E. Rozner, S. Banerjee, P. Bahl, DSAP: a protocol for coordinated spectrum access, in *Proceedings of the IEEE DySPAN*, 2005, pp. 611–614
42. M. Buddhikot, P. Kolodzy, S. Miller, K. Ryan, J. Evans, DIMSUMnet: new directions in wireless networking using coordinated dynamic spectrum, in *Proceedings of the IEEE WoWMoM*, 2005, pp. 78–85
43. Y. Yuan, P. Bahl, R. Chandra, T. Moscibroda, Y. Wu, Allocating dynamic time-spectrum blocks in cognitive radio networks, in *Proceedings of the ACM MobiHoc*, 2007, pp. 130–139

44. L. Zhang, K. Zeng, P. Mohapatra, "Opportunistic spectrum scheduling for mobile cognitive radio networks in white spaces," in *Proceedings of the IEEE WCNC*, 2011, pp. 844–849
45. X. Zhong, Y. Qin, Y. Yang, L. Li, CROR: Coding-aware opportunistic routing in multi-channel cognitive radio networks, in *Proceedings of the IEEE GLOBECOM* 2014, pp. 100–105
46. Network Simulator (ns2). Available at: http://www.isi.edu/nsnam/ns/
47. Michigan Technological University, Cognitive radio cognitive network simulator. Available at http://stuweb.ee.mtu.edu/ljialian/index.htm
48. H. Bany Salameh, M. Krunz, O. Younis, MAC protocol for opportunistic cognitive radio networks with soft guarantees. IEEE Trans Mob Comput **8**(6), 1339–1352 (2009)
49. S. Sengupta, S. Rayanchu, S. Banarjee, An analysis of wireless network coding for unicast sessions: the case for coding-aware routing, in *Proceedings of the IEEE INFOCOM*, 2007, pp. 1028–1036
50. J. Le, J. C. S. Lui, D. M. Chiu, How many packets can we encode? An analysis of practical wireless network coding, in *Proceedings of the IEEE INFOCOM*, 2008, pp. 1040–1048

Chapter 4
Toward a Loss-Free Packet Transmission via Network Coding

Hui Li, Kai Pan, and Shuo-Yen Robert Li

Abstract Network coding promises significant benefits in network performance, especially in lossy environment. As the Transmission Control Protocol (TCP) forms the central part of the Internet protocol, it is necessary to find out the way that makes these benefits compatible with TCP. This chapter introduces a new mechanism for TCP based on network coding which only requires minor changes to the protocol to achieve incremental deployment. The center of the scheme is transmitting linear combination of original packets in the congestion window and simultaneously generating redundant combinations to mask random losses from TCP. Original packets in the congestion window can be deleted even before it is decoded at the receiver side, since the receiver acknowledges the degree of combinations instead of packet itself. Thus, all the original packets can be obtained once enough combinations are collected. Simulation results show the scheme achieves much higher throughput than original TCP in lossy network. Though it still seems far from being deployed in the real network, it has finished the first step in taking the concept of NC into practice.

4.1 Background

History has witnessed the great success of the Transmission Control Protocol (TCP) in the Internet as a connection-oriented protocol tied with the transport layer. Despite the fact that TCP performs very well in the wired networks since its inception, it still suffers a lot in wireless networks due to the external interference and even random loss. The problem stems from its misinterpretation of packet loss which is a sign of congestion in wired networks; thus TCP erroneously reduces the congestion window as a precaution. That is, it slows down the sending rate,

H. Li (✉) • K. Pan
Shenzhen Key Lab of Information Theory and Future networks architecture, Shenzhen Engineering Lab of Converged Networks Technology, School of Electronic & Computer Engineering, Peking University, Beijing, China
e-mail: lih64@pkusz.edu.cn

S.-Y.R. Li
University of Electronic Science and Technology of China, Chengdu, China

© Springer International Publishing Switzerland 2016
Y. Qin (ed.), *Network Coding at Different Layers In Wireless Networks*,
DOI 10.1007/978-3-319-29770-5_4

alleviating congestion and making network stable. Then, the following congestion avoidance phase with a conservative increase in window size causes the transmission link to be underutilized. As more and more wireless technologies are deployed, the problem becomes much sharper when TCP is running over wireless network with high bit error or random packet loss which happens frequently. Thus, wrong interpretation of random loss as the sign of congestion results in TCP suffering performance degradation and is unsatisfactory.

Aiming to the enhancement in wireless network, one promising variant called TCP-Veno [1] was proposed, showing remarkable improvement on performance over legacy TCP. TC-Veno, which concentrates on the non-congestion loss problem, takes advantage of the congestion detection scheme in TCP-Vegas [2] and dexterously combines it with TCP-Reno. In TCP-Veno, two kinds of loss, namely, congestion and random losses, are differentiated by estimating the number of overstocked packets N. Specifically, if the N is smaller than a presetting threshold β, the loss is then considered as random loss, and TCP-Veno only cuts down the congestion window by $\frac{1}{5}$ rather than $\frac{1}{2}$ to maintain the sending rate at a high level. Otherwise, loss is viewed as congestion and the congestion window shrinks to $\frac{1}{2}$ as TCP-Reno does. A key parameter N representing the backlog of packets that are in the current queue is calculated as follows:

$$N = \text{Actual} \times (\text{RTT} - \text{MRTT}) = \left(\frac{\text{cwnd}}{\text{MRTT}} - \frac{\text{cwnd}}{\text{RTT}} \right) \times \text{MRTT}, \qquad (4.1)$$

where MRTT and RTT are the expected and actual RTT, respectively. Although remarkable performance has been achieved by TCP-Veno especially in wireless networks, TCP can still be further improved in many aspects such as redundant packets to mask random losses with initiative.

On referring the redundant packets, traditional TCP can hardly make it since it is impossible to know which packet will be lost in advance. Thus, sending redundant packets blindly not only occupies network resources but also affects the good put. However, things will be different if the redundant packet contains the information of multiple packets which can be realized by coding.

Network coding (NC), emerging as an important promising approach to the transmission of network, means an arbitrary and causal mapping from inputs to outputs of any node in the network. The major advantage of NC is derived from its ability to mix data across time and across flows [3], making data transmission robust and effective. Meanwhile, this technology becomes a potential approach to achieve redundant transmission.

In TCP, the paradigm of orderly delivery with response is one of the key features, which is totally different after coding operation is involved. By collecting enough coded packets, the receiver can recover the original packets regardless of the order they arrived, thus naturally eliminating this kind of limitation on sequence. Along this thought, the sender cares much more about the quantity of the coded packets (degree) that can be received by the receiver instead of which one it received or not. Therefore, a critical issue on how to establish a mapping relationship in the order of disorder transformation should be answered, which provides a basic and feasible way to reform TCP with NC. In short, the scheme of acknowledgment (ACK) should be modified from information response to degree response in order to accommodate the new paradigm.

4.2   Overview

The very first idea of the combination of TCP and NC was proposed in [4] by Sundararajan et al. In fact, the general approach has been to mask losses from TCP using link layer retransmission in [5] before that. However, this kind of scheme is complicated for link layer retransmission to interact with TCP's retransmission which makes performance suffer owing to the independent retransmission protocols at different layers.

On history, there was another similar famous idea known as the digital fountain codes [6] which was used to transmit packets on lossy links. The sender generates a stream of random linear combinations for a batch of k packets, and the receiver can recover the batch as long as it collects slightly more than k combinations. During the transmission, there are no feedbacks except the receiver finished decoding successfully. Although the fountain codes are low in complexity and rate less, its encoding operation has to perform on a batch of packets. That is to say, only after current batch has been received and decoded can the original packet be available. Thus, if this idea is directly used in TCP, either it will lead to timeout retransmission or a very large round-trip time (RTT) which causes low throughput.

Hence, to accommodate the ACK-based protocol, an independent network coding layer should be added below TCP and above IP layer on both sender and receiver sides to process coded packets [5]. Specifically, the additional new layer is dedicated to mask packet loss against upper layer with redundant coded packets (linear combinations). Similar to the fountain codes, once the receiver collects enough combinations, it can recover the original packets via decode operation. This procedure is like solving multivariate equations where the combinations can be considered multivariate unknowns. Compared to the fountain codes, network coding-based scheme generates linear combinations with variable quantity of original packets rather than batch of them to shorten the decoding delay. Here, ACK is sent for each combination as the information degree of freedom rather than information itself. As long as the quantity of equations matches the degree, all the unknowns can be decoded in time. This can explain the mapping relationship referred above.

As one can see from Fig. 4.1, the biggest difference between two kinds of transmission lies in the sending packets. The left one transmits original packets (without NC), while the right one encodes the original packets before sending them out (with NC).When packets are transmitted in the traditional way, the receiver has to obtain definitely every original packet without any losses. Nevertheless, it is difficult to guarantee orderly delivery in wireless networks due to the interference such as high bit error rate or noise in transmission. Consequently, congestion window erroneously shrinks to a small number (caused by RTO or quick retransmission)

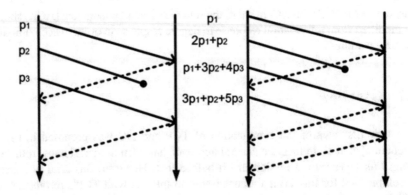

Fig. 4.1 Response of information and degree

and leads to low throughput. If the packets are coded before sending out, the receiver cares much more about the degree of freedom rather than the packet sequence. Thus when a new combination arrives, the receiver will orderly confirm the packets even if the original packets have not been decoded.

4.3 Preliminaries

In order to utilize the standard TCP protocol with minimal change, TCP-Vegas is picked up as it is more compatible, which uses a proactive approach to carry out congestion control by inferring the size of network buffers even before packets get dropped. The crux of the algorithm is to adjust the congestion window and the transmission rate through estimating RTT to find the discrepancy between the expected and actual transmission rate. As congestion arises, buffers start to fill up and the RTT starts to rise, which can be used to adjust the congestion window and the rate as a congestion sign [2]. Due to the RTT-aware characteristic of Vegas, the congestion window will not be affected too much but usually causes the latest RTT a little larger if a packet is lost. Comparatively, TCP-Reno which uses packet loss as a congestion indicator and shrinks congestion window drastically by half when loss packet is detected may not be so suitable to work with NC.

Before being applied NC, the message is split into a stream of packets $P_1, P_2, \cdots$. The packet is treated as a vector over a finite field F_q of size q. Thus in the NC system, a node also performs a linear network coding to incoming packets in addition to forwarding them. That means the node may transmit a packet obtained by linearly combining the vectors corresponding to the incoming packets. For instance, it may transmit $q_1 = \alpha p_1 + \beta p_2$ and $q_2 = \gamma p_1 + \delta p_2$, where $\alpha, \beta, \gamma, \delta \in F_q$ are the coefficients. Assuming the packets have l symbols, the encoding process may be written in matrix form as

$$\begin{pmatrix} q_{11} \, q_{12} \cdots q_{1l} \\ q_{21} \, q_{22} \cdots q_{2l} \end{pmatrix} = C \cdot \begin{pmatrix} p_{11} \, p_{12} \cdots p_{1l} \\ p_{21} \, p_{22} \cdots p_{2l} \end{pmatrix}, \tag{4.2}$$

where $C = \begin{pmatrix} \alpha & \beta \\ \gamma & \delta \end{pmatrix}$ is the coefficient matrix. Therefore, original packets can be obtained by inverting matrix C using Gauss-Jordan elimination after receiving q_1 and q_2, which is given by

$$\begin{pmatrix} p_{11} \ p_{12} \ \cdots \ p_{1l} \\ p_{21} \ p_{22} \ \cdots \ p_{2l} \end{pmatrix} = C^{-1} \cdot \begin{pmatrix} q_{11} \ q_{12} \ \cdots \ q_{1l} \\ q_{21} \ q_{22} \ \cdots \ q_{2l} \end{pmatrix}, \tag{4.3}$$

As a result, the receiver needs to receive as many linear combinations as the number of original packets involved to recover all of them.

In the above setting, some definitions are introduced [7] that will be useful throughout this chapter.

Definition 1. Seeing a packet: A node is said to have seen a packet p_k if it has enough information to compute a linear combination of the form $(p_k + q)$, where $q = \sum_{l>k} \alpha_l p_l$, with $\alpha_l \in F_q$ for all $l > k$. Thus, q is a linear combination with packet indices larger than k. Actually, "seen packets" reflect the ability of decoding packets. When seen equals the number of coded original packets, original packets can be recovered via decoding operation. Algebraically speaking, the procedure is like solving linear equations, which explains the fact that orderly confirmation of original TCP does not exist anymore.

Definition 2. Knowledge of a node: The knowledge of a node is the set of all linear combinations of original packets. It equals the total number of original packets involved in the combinations or the degree of freedom. If a node has seen packet p_k, then it knows exactly one linear combination of the form $p_k + q$, where q is the vector which includes only unseen packets in terms of Definition 1.

4.4   Network Coding-Based TCP

As a solution to mask losses from TCP layer, the protocol is needed to be modified to suit accommodating NC function. There are mainly three categories of schemes to implement NC on TCP, each of which has a same architecture as shown in Fig. 4.2 where new TCP coexists with the original TCP.

4.4.1   Static Methodology

As the seminal work of NC-based TCP [4], the primary aim is to mask losses from TCP using random linear coding. Thus the first problem is how to acknowledge the data. According to the definitions in the last section, the degree of freedom is more important in the new protocol and the notion of seen packets also achieves an

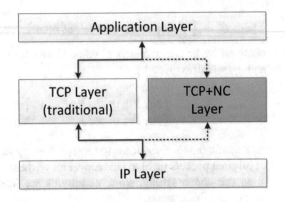

Fig. 4.2 Illustration of the new TCP layer

ordering of the degree of freedom. Thus, once receiving a linear combination, the node finds out the new seen packet if there is any.

The idea of masking losses lies in the fact that every new arrival random linear combination will cause the next unseen packet to be seen with high probability. Therefore, once the number of combinations reaches the value of the degree of freedom, the node can decode all the original packets and deliver them to the upper layer. Given the loss rate, the number of combinations sent by the source node should be slightly larger than the number of original packets which is controlled by the redundancy parameter R. The redundancy is the ratio of actual sending packets and least sending packets (without loss) which is preset. It is notable that the redundancy should be set neither too small nor too large. If there is too little redundancy, the losses are not effectively masked from TCP layer which will lead to low throughput. On the other extreme, it is also bad, not only because the transmission rate becomes limited by the rate of code but also because too many packets may congest the network. Hence the ideal R should be kept equal to the reciprocal of the probability of successful reception in theory. However, in the static strategy, redundancy keeps the same once it was set during the transmission. The algorithm is specified as follows in the NC module, separating into source node and receiver.

Source node: Respond to the arrival of a packet from source node and an ACK from the receiver.

1 $NUM \leftarrow 0$;
2 wait until a packet comes;
3 **if** (control packet)
4 deliver to the IP layer and return to 2;
5 **if** (ACK from the receiver)
6 remove the ACKed packet from the coding window and deliver the ACK to the source node;
7 **if** (new data packet)
8 add to the coding window;
9 $NUM \leftarrow NUM + R$ (R is redundancy);

10 **while** ($i \leq \lfloor NUM \rfloor$)
11 generate a random linear combination of the packets in the coding window;
12 add the coefficients to the network coding header and deliver to the IP layer;
13 $NUM \leftarrow NUM - \lfloor NUM \rfloor$;
14 return to 2.

Packets used for the connection management or acknowledgment are directly forwarded (lines 3–6). For the new data packets, they are first added to the coding window in NC module and then generated linear combinations according to NUM with coefficients added to the NC header (lines 7–12). Finally NUM is updated by the fraction of current NUM for the next time (line 13).

Receiver: Respond to the arrival of a packet from source node and an ACK from the receiver.
1 wait until a packet comes;
2 **if** (ACK from receiver)
3 **if** (control packet)
4 deliver to the IP layer and return to 1;
5 **else**
6 ignore and return to 1;
7 **if** (combinations from source node)
8 retrieve the coefficients and perform Gauss-Jordan elimination to update the seen packets;
9 perform corresponding operations to the payload as the coefficients and deliver the decoded packets to the upper layer;
10 generate an ACK with the next unseen packet;
11 return to 1.

Similar to the source node, control packets are directly forwarded (lines 3–4). For the new arrival combination, Gauss-Jordan elimination is performed on the coefficients to update the seen packets, and same operations are done to the payload if the combination causes a new packet to be seen (lines 7–9). After that, an ACK with the latest seen packet or the next unseen packet is sent to the source node (line 10).

The soundness of the new protocol can be guaranteed since every packet will be delivered reliably to the receiver eventually. Though the ACK mechanism is a little different from the conventional one which removes a packet once it is received, every packet can be decoded as long as they have been seen. To make the congestion control scheme work, whenever a new packet enters the congestion window, it is transmitted to the NC module. Packets transmitted earlier that have not been decoded can be removed from the coding window. This will not be a problem since redundant packets can eventually make every packet be seen and decoded.

Static scheme is effective only if the redundancy approximates or exceeds the theoretical value, and thus it is suitable for the environment with quasi-constant loss rate.

4.4.2   Dynamic Methodology

To make the new protocol available in the network without constant loss rate, Chen et al. proposed a dynamic scheme in [8]. In the beginning, the scheme was proposed to decrease the decoding delay, since the static scheme does not consider this issue and has to buffer a considerable number of packets before decoding all packets. Therefore, it is hard to deploy in a real system in spite of the benefits brought by NC besides the unadjustable redundancy.

Dynamic scheme retransmits the appropriate number of random linear combinations in terms of the feedback information when loss happens. The appropriate number is decided by the difference between the largest packet index *max_index* in the combination and the largest seen packet index *max_seen*. Thus, the difference termed *loss* mentioned above is embedded in the ACK in conjunction with the sequence number of the largest seen packet *max_index*. For instance, consider source node sent three random linear combinations $x = p_1$, $y = 2p_1 + 3p_2$, and $z = p_1 + 3p_2 + 4p_3$ to the receiver with the second combination y lost. The source node will receive an ACK with max $index = 3$ and $loss = 3 - 2 = 1$ after combination z reached the receiver. Here, *loss* indicates the number of packets still needed by the receiver to decode all the combinations.

Dynamic scheme uses the *loss* to decide the number of retransmission combinations instead of redundancy parameter R. The retransmission algorithm is specified as follows:

Algorithm of retransmission.
  1 wait until an ACK comes;
  2 **if** *(Tnow − Tlast > RTO)*
  3 retransmit *loss* linear combinations of first *loss* packets in the coding window;
  4 **else**
  5 **if** *(loss > lastloss)*
  6 retransmit *loss − lastloss* combinations of first *loss − lastloss* packets in the coding window;

Here, Tnow is the current time, Tlast is the time of the last retransmission, and RTO is the timeout value set by TCP. Similarly, *last_loss* is the loss value in the last time. The algorithm first checks to see whether timeout happens since the last retransmission combinations were sent out (line 2). If timeout happens, considerate represents the last retransmission is lost and new loss combinations (line 3) are retransmitted. If there are extra lost combinations after the last retransmission, *loss − lastloss* combinations of first *loss − lastloss* packets instead of all the packets are sent out to decrease the decoding delay (lines 4–6). Compared to the static scheme, dynamic scheme can handle both random losses and unknown bursty losses in time. As a result, decoding delay can be reduced.

4.4.3   Adaptive Methodology

Static scheme masks random loss through accumulating the fraction of current R in prior to the next transmission, while dynamic scheme can detect the new lost packets once an ACK is received by the source node and retransmits new combinations immediately if necessary. Thus the static scheme can be also considered as a kind of preventive scheme. For a lossy environment especially wireless network, chances are that the retransmit combinations will also be lost. However, dynamic scheme can hardly solve this problem until retransmission timeout, since the value of $loss - lastloss$ stays the same during this period. Therefore, it is necessary to collect the advantages of both schemes to completely exploit the superiority of NC.

Adaptive scheme [9] combines both advantages of static and dynamic scheme. Moreover, it makes redundancy adaptive to the environment rather than constant during the transmission. In the scheme, the redundancy is calculated and updated by source node once it receives the feedback. The adjusting algorithm is specified as follows:

Algorithm of redundancy adjustment.
    1 wait until an ACK comes;
    1 **if** $(loss - lastloss < 0)$
    2 $R \leftarrow R + (loss - lastloss) \times \frac{pktsize}{B \times D}$ and return to 1;
    3 **if** $(R < 1)$
    4 $R \leftarrow 1$;
    5 **if** $(loss - lastloss = 0)$
    6 return to 1;
    7 **if** $(loss - lastloss > 0)$
    8 retransmit $loss - lastloss$ combinations of first $loss - lastloss$ packets in the coding window;
    9 $R \leftarrow R + (loss - lastloss) \times \frac{pktsize}{B \times D}$ and return to 1.

Here, B and D are the link bandwidth and link transmission delay, respectively. If $loss - lastloss$ is smaller than zero, it represents one or more retransmission packets have already been received by the receiver and no more combinations should be sent this time. If $loss - lastloss$ is positive, extra combinations are needed. $\frac{B \times D}{pktsize}$ is the number of packets transmitted in the direction from source node to sink node. Thus the function of $(loss - lastloss) \times \frac{pktsize}{B \times D}$ is to send the retransmission packet again quickly when the last retransmission is lost, since the ACK of the retransmission packet should be back before this moment. In adaptive scheme, R reflects the situation of underlying links RTT/2 ago and launches another retransmission after several RTTs which could make at least a part that can be decoded.

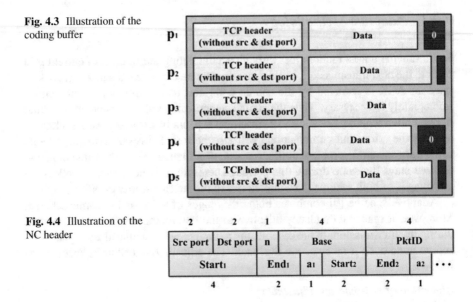

Fig. 4.3 Illustration of the coding buffer

Fig. 4.4 Illustration of the NC header

4.5 Packet Format

Packet format should be redesigned to support NC function which contains coefficient, redundancy, and so on. In the new protocol, an entire packet serves as the basic unit of data, that is, the exact same operation is being performed on every symbol within the packet. The main advantage of this is to perform decoding matrix operations at the granularity of packets, and one coefficient can be used for one packet instead of each symbol which is typically one-byte long. TCP may generate segments in different sizes which are controlled by maximum transmission unit. The solution to the variable length and repacketization in [7] is as follows:

- Any part of the incoming segment already in the buffer is removed from the segment.
- Remaining contiguous part of the segment is repacked in a new separate TCP packet.
- The source and destination ports are removed to be added in the NC header.
- Packets are appended with zeros to achieve the same length as the longest packet in the coding buffer as shown in Fig. 4.3.

Here, the reason the ports are removed from original TCP header is that they are used to identify which TCP connection the packet corresponds to. Thus, they are added to the combination as a part of NC header.

As per the NC header, it is also designed in detail in [7] and some modification is made to work with the improved algorithm as in Fig. 4.4.

The meaning of each field is described below with size (in bytes) written around.

- *Src and dst port*: Used to identify which TCP connection the coded packet corresponds to

- n: The number of packets involved in the combination
- *Base*: The sequence number of the first unacknowledged byte
- *PktID*: The ID of the combination
- $Start_i$: The starting byte of the data in ith packet
- End_i: The last byte of the data in ith packet
- a_i: The coefficient of the ith packet in the combination

The $Start_i$ and End_i are relative to the previous packet's corresponding field except $Start_1$. *PktID* is used to calculate the lost packets and update new seen packets in conjunction with n. *Base* is used to decide whether the packet can be deleted from the buffer safely which is the oldest byte in the coding buffer. Thus the NC header is $15 + 5n$ bytes including the port information.

Another important concept that should be introduced is the coding window, which decides the number of packets in a combination. Theoretically, the sender generates a random linear combination of all the packets in the coding buffer. However, it may result as a huge size of NC header. Thus, a constant number of the packets are coded together and the number depends on the loss rate of the network environment. Because only if the difference between the number of redundant packets received and the number of original packets involved is less than the coding window, the loss can be masked from TCP.

4.6   Experiment

The topology used in the experiment is shown in Fig. 4.5. It consists of several sources S_i, $(i \in 1, \cdots, n)$, several receivers R_i, $(i \in 1, \cdots, n)$, and three intermediates M_i, $(i \in 1, \cdots 3)$. Each packet starts transmitting from S_i, with R_i as the intended destination. The channel between each pair of nodes is assumed to be independent of each other.

4.6.1   Soundness

The new protocol is based on TCP-Vegas since TCP-Vegas controls the congestion window more smoothly via using RTT, compared to the more abrupt manner in

Fig. 4.5 Illustration of the topology in experiment

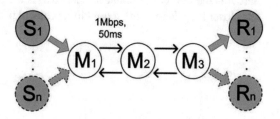

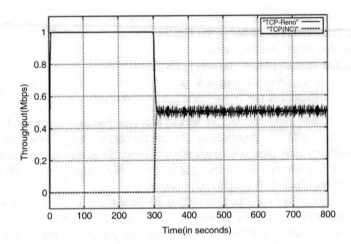

Fig. 4.6 Illustration of the fairness between two flows

TCP-Reno. However, if the coding window is properly chosen, TCP-Reno can also work with NC. According to [7], the value of coding window should be large enough to mask the link loss as well as small enough to make the queue drops visible to TCP which is a little complex to fulfill.

Despite that research has shown the superiority of TCP-Vegas over TCP-Reno [10], TCP-Reno is still the most widely deployed variant of TCP. When both types of the connections share a link, TCP-Reno generally steals bandwidth from TCP-Vegas and dominates due to its aggressive nature, compared to the more conservative mechanism of TCP-Vegas [11]. However, they can be compatible with one another if TCP-Vegas is configured properly [12]. Here, the parameters of TCP-Vegas are set to be $\alpha = 28$, $\beta = 30$, and $\gamma = 2$.

The result of soundness for one flow of TCP-Reno and the one of TCP with NC function is shown in Fig. 4.6. Here the loss rate is set to 0 % and two flows join in the network one after another.

As per the situation of more flows coexist in the network, Fig. 4.7 shows the result, where Jain's fairness index [13] is used with definition as follows:

$$f = \frac{\left(\sum_{i=1}^{n} x_i\right)^2}{n \sum_{i=1}^{n} x_i^2}, \tag{4.4}$$

Here n is the number of connections and x_i is the throughput of the ith connection. The closer the f is to 1, the more fair a protocol becomes.

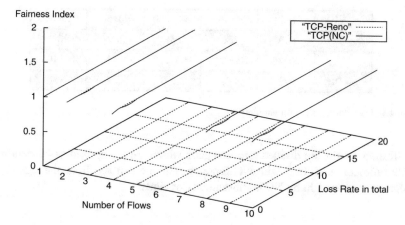

Fig. 4.7 Illustration of the fairness among multiple flows [9]

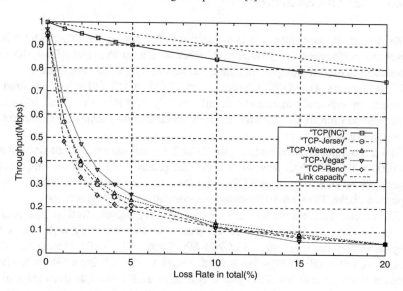

Fig. 4.8 Illustration of the fairness among multiple flows [9]

4.6.2   Effectiveness

Random loss rate is configured on each link to show the effectiveness with NC function. Suppose the random loss rate on each link is p and the overall probability of packet loss is $1 - (1 - p)^{n-1}$, where n is the number of intermediate nodes. The loss rate in experiment varies from 0 to 20 % in which flow with different protocols is tested as in Fig. 4.8.

Time (in seconds)	1~20	20~100	100~250	250~500	500~550
Loss Rate	10%	5%	8%	12%	20%
Time (in seconds)	550~700	700~800	800~900	900~950	950~1000
Loss Rate	10%	5%	10%	20%	5%

Fig. 4.9 Loss rate vs. time for simulation

Testing results are divided into two parts which are recognized as conventional TCP protocols and TCP with NC function. As one can see from the figure, the enhancement brought by NC is substantially obvious.

4.6.3 Network Switching

In the last subsection, loss rate is a constant value during the experiment which makes the performance of different NC schemes almost the same. Thus, the next experiment is used to compare the three NC schemes mentioned in Sect. 4.4. Here, TCP-NC (static), TCP-DNC (dynamic), and TCP-NCDR (adaptive) are used to represent the scheme, respectively. Simultaneously, loss rate is changed drastically from time to time to simulate the network switching in practical environment as shown in Fig. 4.9.

The throughput is calculated at intervals of 2.5 s, and the experiment is simulated for 1000 s. For TCP-NC, R is set to 1.087 (reciprocal of 92 %) as the initial value which can mask losses from TCP if the loss rate is less than or equal 8 % as explained above. Figure 4.10 illustrates the comparison of three schemes.

As one can see from the figure, TCP-NC performs equally well as TCP-NCDR during the time 20–250 s, since R can match the loss rate very well. The same happens during the time interval of 700–800 s and 950–1000 s. During the other time, R is too small to effectively mask losses from TCP layer which leads to frequent timeouts and low throughput. Although TCP-DNC has obtained a good performance both in throughput and robustness, there is a further improvement with TCP-NCDR. In this scenario, continuous random loss is the dominant mode during the stable phase, while intermittent random loss plays an important role in saltus phase.

The design of TCP-DNC focuses on the difference between neighboring losses which neglects the accumulated redundant packets against successional random loss. Given redundant packets may also lose, it will not be sent out again unless another saltus of loss or timeout. Consequently, congestion window shrinks slowly by interpreting this phenomenon as a sign of congestion under TCP-DNC which is based on TCP-Vegas until packets are decoded. Thus, it performs slightly poor during the stable phase compared to TCP-NCDR. Despite the fact that adaptive scheme outperforms the other two, it requires more feedback and calculation

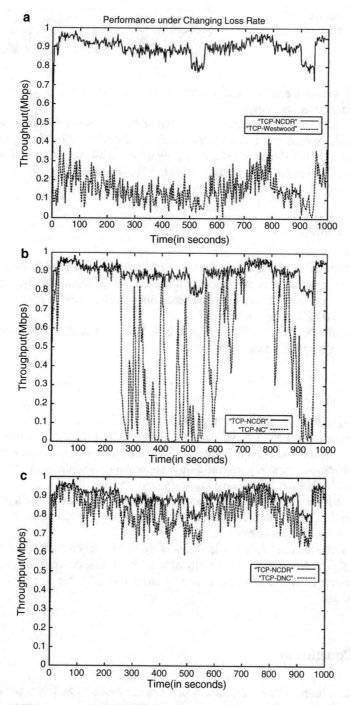

Fig. 4.10 (**a**) Throughput vs. time for different schemes (TCP-Westwood) [9]. (**b**) Throughput vs. time for different schemes (TCP-NC) [9]. (**c**) Throughput vs. time for different schemes (TCP-DNC) [9]

Fig. 4.11 Average
throughput of different
schemes

Westwood	TCP-NC	TCP-DNC	TCP-NCDR
0.147Mbps	0.58Mbps	0.697Mbps	0.843Mbps

overhead during the transmission which is possible to reduce this overhead by further optimizing the scheme.

Figure 4.11 shows the average throughput of each scheme during the switching experiment. As one can see from the result, TCP-NCDR outperforms the other scheme due to the dynamic redundancy, while conventional TCP such as TCP-Westwood [14] only achieves a small amount of throughput.

4.6.4   Multi-flows

This subsection considers the performance improvement when multi-flow joins in and the topology used in the experiment is the same as shown in Fig. 4.5. Here, random loss rate p is also configured on each link which makes the general loss rate varies from 0 to 20 % as the above.

The number of flows in Fig. 4.12 is 1, 2, 4, 8, and 10, respectively. Throughput is defined as normalized one as follows:

$$T = \frac{T_{NC}}{T_{Reno}}, \tag{4.5}$$

where T_{NC} and T_{Reno} are the average throughput of NC scheme and conventional TCP, respectively. Here the NC scheme represents TCP-NCDR. As one can see from each figure, NC scheme outperforms conventional TCP especially under the lossy environment, and it performs a little better than conventional TCP when the load becomes heavier in a relative low lossy environment.

For the following experiment, a small modification is done to the topology as shown in Fig. 4.13 where the channel from intermediate node to the receivers is substituted by wireless channel.

As one can see from the result in Fig. 4.14, NC-based scheme still achieves better in throughput. However, as the quantity of flows increases, the collision probability also increases, making throughput improvement less effective than in the wired environment.

4.7   Conclusion

Up to now, there is rarely a protocol that can substitute TCP. It has been proven significantly successful in the Internet as a connection-oriented protocol tied with

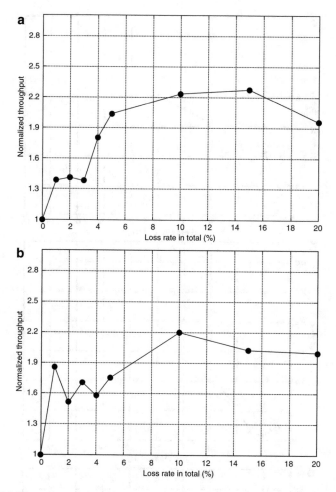

Fig. 4.12 (a) Illustration of the effectiveness of multi-flows in wired network (one flow). (b) Illustration of the effectiveness of multi-flows in wired network (two flows). (c) Illustration of the effectiveness of multi-flows in wired network (four flows). (d) Illustration of the effectiveness of multi-flows in wired network (eight flows). (e) Illustration of the effectiveness of multi-flows in wired network (ten flows)

the transport layer. However, it is reasonable that conventional TCP becomes increasingly mismatched to the transmission requirements. Thus, network coding, which has the ability to mix data across time and across flows, may become an important potential approach to tackle with this problem. This chapter briefly introduces the combination of TCP and NC, from idea to preliminary simulation, which demonstrates a robust and effective improvement in data transmission especially in lossy environment. Despite the notable results in the experiment, it still seems far from being deployed in the practical network. The current work can

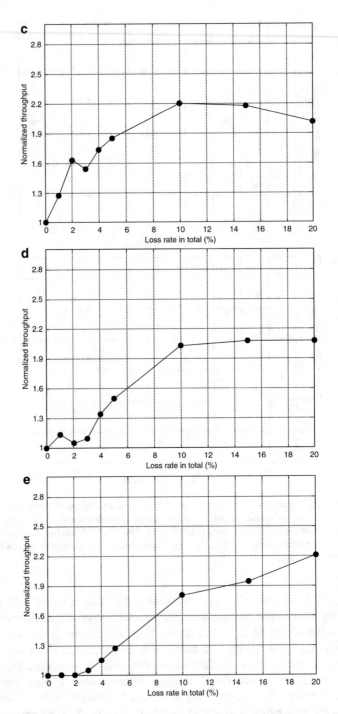

Fig. 4.12 (continued)

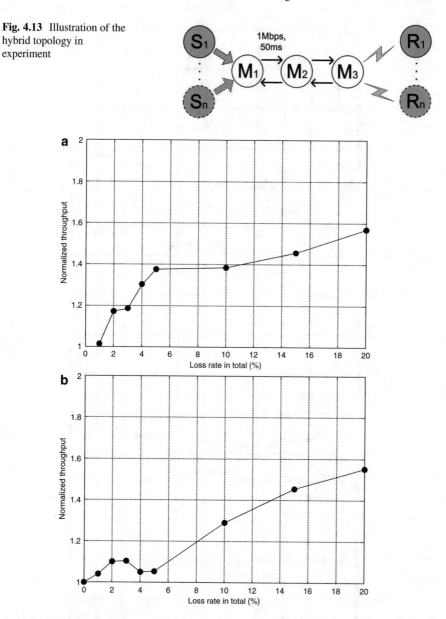

Fig. 4.13 Illustration of the hybrid topology in experiment

Fig. 4.14 (**a**) Illustration of the effectiveness of multi-flows in hybrid network (one flow). (**b**) Illustration of the effectiveness of multi-flows in hybrid network (two flows). (**c**) Illustration of the effectiveness of multi-flows in hybrid network (four flows). (**d**) Illustration of the effectiveness of multi-flows in hybrid network (eight flows). (**e**) Illustration of the effectiveness of multi-flows in hybrid network (ten flows)

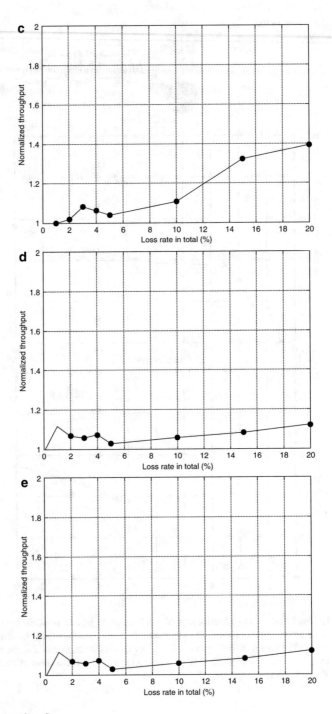

Fig. 4.14 (continued)

be viewed as a first step in taking the concept of NC into practice, and many left-open questions such as re-encoding at intermediate nodes in conjunction with the estimation on calculation overhead will be researched intensively in the near future.

References

1. C.P. Fu, S.C. Liew, TCP veno: TCP enhancement for transmission over wireless access networks. IEEE J Select Areas Commun **21**(2), 216–228 (2003)
2. L. S. Bramko, S. W. O'Malley, L. L. Peterson, TCP Vegas: new techniques for congestion detection and avoidance, in *Proceedings of ACM SIGCOMM Symposium*, Aug 1994, pp. 24–35
3. T. Ho, Networking from a Network Coding Perspective, Ph.D Thesis, Massachusetts Institute of Technology, Dept. of EECS, May 2004
4. J. K. Sundararajan, D. Shah, M. Medard, M. Mitzenmacher, J. Barros, Network coding meets TCP, in *Proceedings of IEEE INFOCOM*, Apr 2009, pp. 280–288
5. S. Paul, E. Ayanoglu, T. F. L. Porta, K.-W. H. Chen, K. E. Sabnani, R. D. Gitlin, An asymmetric protocol for digital cellular communications, in *Proceedings of IEEE INFOCOM*, Apr 1995, pp. 1053
6. M. Luby, LT Codes, in *Proceedings of IEEE Symposium on Foundations of Computer Science*, Oct. 2002, pp. 271–280
7. J.K. Sundararajan, D. Shah, M. Medard, M. Mitzenmacher, J. Barros, Network coding meets TCP: theory and implementation. Proc IEEE **99**, 490–512 (2011)
8. J. Chen, W. Tan, L. Liu, X. Hu, F. Xu, Towards zero loss for TCP in wireless networks, in *Proceedings of IEEE IPCCC*, Dec 2009, pp. 65–70
9. K. Pan, H. Li, S. Y. Li, Q, Shi, W. Yin, Flying against lossy light-load hybrid networks, in *Proceedings of IEEE IWCMC*, Apr 2013, pp. 252–257
10. P. A. Chou, Y. N. Wu, K. Jain, Practical network coding, in *Proceedings of Allerton Conference on CCC*, 2003
11. W. Feng, S. Vanichpun, Enabling compatibility between TCP Reno and TCP Vegas, in *Proceedings of IEEE Symposium on Applications and the Internet*, Jan. 2003, pp. 301–308
12. L. S. Bramko, S. W. O'Malley, L. L. Peterson, TCP Vegas: new techniques for congestion detection and avoidance, in *Proceedings of SIGCOMM*, Aug. 1994, pp. 24–35.
13. R. Jain, *The art of computer system performance analysis* (Wiley, New York, 1991)
14. C. Casetti, M. Gerla, S. Mascolo, M.Y. Sanadidi, R. Wang, TCP west wood: end-to-end congestion control for wired/wireless networks. J Wireless Netw **8**(5), 467–479 (2002)

Chapter 5
Network Coding in Application Layer Multicast

Min Yang and Yuanyuan Yang

Abstract It is proved that network coding can achieve *multicast capacity* and therefore improves the throughput of a multicast network significantly. This chapter will focus on applying network coding to *Application layer multicast*(ALM). The benefits are two folds: first, ALM is built on peer-to-peer networks whose topology can be arbitrary so it is easy to tailor the topology to facilitate network coding; second, the nodes in ALM are end hosts which are powerful enough to perform complex encoding and decoding operations.

5.1 Background

5.1.1 Peer-to-Peer and ALM

The emerge of *peer-to-peer* technology provides a promising alternative solution for multicast communication [37, 39, 40, 46]. Peer-to-peer is referred to as a fully distributed network architecture which is in contrary to the traditional server–client model. In server–client model, a server is providing centralized service requested by different clients, i.e., hosts.[1] Usually the address of the server is well known by the hosts in advance. The most obvious drawback of server–client model is its limited bandwidth and resource on the server side. Since the bandwidth and resource of the server is shared by all the hosts, the server can be easily overwhelmed by a huge number of simultaneous hosts. Peer-to-peer model allows hosts to form an adhoc logical overlay network on top of the physical network. The links of the overlay network are logical links each of which can be mapped to a physical path

---

[1]In this chapter, host is used to represent the end users of the Internet. Every host is connected to a router to access the Internet.

M. Yang (✉)
Google Inc., Mountain View, CA 94043, USA
e-mail: minyang@google.com

Y. Yang
Stony Brook University, Stony Brook, NY 11794, USA

© Springer International Publishing Switzerland 2016
Y. Qin (ed.), *Network Coding at Different Layers In Wireless Networks*,
DOI 10.1007/978-3-319-29770-5_5

in the physical network. Hosts can share information as well as bandwidth with each other through the overlay network. This requires hosts be able to perform more complex operations such as routing and overlay topology construction/maintenance. The hosts are called peers as they are equivalent in terms of their functionalities. The advantages of peer-to-peer systems are obvious. First, the larger the system size is, the larger the total bandwidth is. As peers can share their bandwidth with each other, peers can contribute their bandwidth to the system. More peers join the system, more bandwidth the system has. Second, the processing is distributed, no *single point of failure* problem. In server–client model, the server is much more important than the hosts. If the server is down, the whole system is down as well. This is called single point of failure problem. In peer-to-peer systems, peers are of equal importance. The overlay network is formed in a distributed way such that the system can continue to work properly even if some peers leave the system.

In practice, a lot of peer-to-peer systems [3, 24, 43] adopt a hybrid model. There is still a server holding the resource which is requested by a lot of hosts. To make the system scalable, the hosts form a peer-to-peer network and help each other to retrieve the resource. As a result, the server can serve much more hosts than that in server–client model.

Application layer multicast(ALM) [31] is proposed to circumvent multicast support in routers by implementing multicast related functionalities in hosts' application layer. A peer-to-peer network is formed between the source and all the receivers. Then a multicast tree is constructed over the peer-to-peer network. Similarly, the messages are transmitted from the root to the leaves. On the physical network level, the messages are transmitted through unicast along the paths indicated by the tree. As traditional multicast routing implements multicast support in network layer, it is often referred as *network layer multicast*. ALM is a promising alternative for multicast communication over a large scale network. It does not need router support, so it is easy to deploy. It can support infinite receivers in theory. Since receivers help forwarding messages for each other, the more the receivers, the higher the total uploading bandwidth.

Although some materials covered in this chapter are presented in scenarios of wired networks, they apply to both wireless and wired networks.

5.1.2 *Network Coding for Multicast Networks*

Network coding is proposed recently as a generalization of routing [22]. Routing allows relay nodes to forward or duplicate messages. While network coding allows relay nodes to encode messages. Forwarding or duplicating is considered as a special case of encoding. Network coding has a lot of potential applications for both wired and wireless networks [9, 30, 41, 42, 47, 50]. Multicast is one of the most important application of network coding. In [22], the authors prove that with network coding, a multicast network can achieve its maximum throughput.

Fig. 5.1 Illustration of
network coding advantage
over routing. **(a)** Multicast
without network coding
(b) Multicast with network
coding

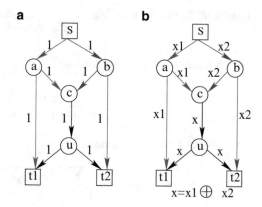

First here is an example to illustrate the advantage of network coding over routing
to a multicast network. As shown in Fig. 5.1, node s is the source node, $t1$ and $t2$
are two receivers. All edges in the figure have capacity 1, which means the edge can
only transmit 1 unit data (bit) per unit time (sec). Source s has two bits, $x1$ and $x2$ to
multicast to both $t1$ and $t2$. Traditional multicast without network coding is shown
in Fig. 5.1a. Without loss of generality, bit $x1$ is represented by the red flow, bit $x2$
is represented by the blue flow, and both bits $x1$ and $x2$ are represented by the black
flow. Bit $x1$ can reach $t1$ in 2 s. Bit $x2$ can reach $t2$ in 2 s. When node c receives both
bits, it forwards them in sequence. Suppose it forwards bit $x1$ first. Then $t1$ receives
both bits in 4 s and $t2$ receives both bits in 5 s. Now consider using network coding
on link cu. When node c receives both bits, it first mixes them by operation XOR.
Then it sends the mixed bit to node u. When node $t1$ or $t2$ receives the mixed bit x, it
can recover the original bits $x1$ and $x2$ by XOR the mixed bit and the other received
bit. All the transmission can be completed in 4 s. Therefore the throughput of the
network can be increased by $\frac{1}{3}$ using a very simple network coding.

Based on the type of the encoding function at the relay nodes, network coding can
be categorized into two types: linear network coding, where the encoding functions
are linear functions and non-linear network coding, where the encoding functions
are non-linear functions. Li et al. proved [48] that linear network coding is sufficient
for a multicast network to achieve its maximum throughput.

5.1.3 Linear Network Coding for Multicast

Here are some preliminary theories about linear network coding for multicast.
A network can be represented as a directed graph $G(V, E, C)$ where V is the set
of nodes and E is the set of edges and C is the link capacity function, $C : E \rightarrow \Re^+$.
Suppose s is the source, $t_1, \ldots, t_L$ are L receivers. Here are the definitions of two
terms: *flow* and *cut*.

Definition 1 (Flow). $F = [F_{ij}, (i,j) \in E]$ is a *flow* in G from s to t_l if for all $(i,j) \in E$

$$0 \leq F_{ij} \leq C_{ij}$$

such that for all $i \in V$ except for s and t_l

$$\sum_{i':(i',i)\in E} F_{i'i} = \sum_{j:(i,j)\in E} F_{ij}$$

The value of flow F is defined as

$$\sum_{j:(s,j)\in E} F_{sj} - \sum_{i:(i,s)\in E} F_{is}$$

which is equal to

$$\sum_{i:(i,t_l)\in E} F_{it_l} - \sum_{j:(t_l,j)\in E} F_{t_l j}$$

F is a *max-flow* from s to t_l in G if F is a flow from s to t_l whose value is greater than or equal to any other flow from s to t_l. The max-flow F is the upper bound of the transmission rate between s and t_l.

Definition 2 (Cut). A *cut* in G between s and t_l means a collection of H of nodes which includes s but not t_l. An edge ij is said to be in the cut H if $i \in H$ and $j \in \bar{H}$. The value of the cut is the sum of the capacity of all the edges from H to $\bar{H}$.

Theorem 1 (Max-Flow Min-Cut Theorem). *For every nonsource node t, the minimum value of a cut between the source and the node t is equal to the max-flow between the source and t.*

The Max-Flow Min-Cut Theorem provides a simple way to calculate the max-flow by finding the minimum cut.

In a unicast session, the maximum transmission rate between the source and the receiver is the max-flow between them. In a multicast session, *multicast capacity* is defined as the minimum of max-flows between the source and the receivers. If the source must transmit messages to all the receivers at the same rate, it is easy to see that the multicast capacity is the upper bound of the transmission rate. Therefore multicast capacity is considered as the maximum throughput a multicast network can achieve.

In their pioneer paper [22], Ahlswede et al. prove that with network coding, the maximum throughput can be achieved for a multicast network, i.e., network coding can take full advantage of multicast capacity. In paper [48], Li et al. prove that linear network coding is sufficient to achieve this goal and even a stronger result: with linear network coding, each receiver can receive the messages at the rate determined by the max-flow between the source and the receiver simultaneously. In paper [44], a new novel linear network coding construction scheme is discussed.

Here is an example to explain the notion of linear network coding. Given a graph
$G(V,E)$, each node has multiple incoming edges and outgoing edges. An edge e is
represented by an ordered node pair (x,y), $x,y \in V$, where y is called the head of the
edge and denoted as $y = \text{head}(e)$, and x is called the tail of the edge and denoted as
$x = \text{tail}(e)$. The messages can only be transmitted from x to y. The incoming edge
set of a node v is defined as

$$E_{\text{in}}(v) = \{(x,y)|(x,y) \in E, y = v\}. \tag{5.1}$$

Similarly, the outgoing edge set of a node v is defined as

$$E_{\text{out}}(v) = \{(x,y)|(x,y) \in E, x = v\}. \tag{5.2}$$

Assume each edge $e \in E$ can carry one symbol from a certain finite field F. Let y_e
denote the symbol carried by edge e. Let $\mathbf{x} = (x_1,\ldots,x_h)$ denote the source symbols
available at source s. For notational consistency, h source edges are introduced,
$s_1,\ldots,s_h$, which all end in s; the source edges $s_1,\ldots,s_h$ carry the h source symbols
$x_1,\ldots,x_h$, respectively. Since each node encodes the incoming messages with a
linear function, the symbol on an outgoing edge is a linear combination of the
symbols on the incoming edges, namely

$$y_e = \sum_{e':\text{head}(e')=\text{tail}(e)} w_{e'e} y_{e'}. \tag{5.3}$$

The coefficients $W = w_{e'e}$ are called *mixing coefficients*. The set of all mixing
coefficients can be collectively represented by a $|E| \times (|E|+h)$ matrix. The structural
restriction of W is that $w_{e'e} = 0$ unless $\text{head}(e') = \text{tail}(e)$. Partition W into two
parts as

$$W_{|E|\times(|E|+h)} = [A_{|E|\times|E|}\ B_{|E|\times h}], \tag{5.4}$$

where the subscript indicates the size of the matrix.

Equation (5.4) can be rewritten in a matrix form as

$$\mathbf{y} = A\mathbf{y} + B\mathbf{x}. \tag{5.5}$$

Suppose the graph is acyclic,[2] we topologically sort the edges. Then the matrix A
is a lower triangular matrix with zeros on the diagonal line. Thus (I-A) is invertible
and

$$\mathbf{y} = (I - A)^{-1}B\mathbf{x}. \tag{5.6}$$

---

[2]If the graph is cyclic, a subgraph which satisfies our requirement can be found.

Therefore, y_e on any edge e is a linear combination of the source symbols. Define $Q_{|E| \times h} \equiv (I - A)^{-1}B$. The eth row of Q is denoted as q_e, which is called the *global encoding vector* at edge e since $y_e = q_e x$.

Since each y_e is a linear combination of the source symbols, any receiver t receiving h symbols with linearly independent global coding vectors can decode the source symbols by solving the corresponding system of linear equations.

The above discussion is summarized in the following definition.

Definition 3. Linear Network Coding Assignment: Given an acyclic graph G(V,E) a source node s, a finite field F, and a code dimension h, a linear network coding assignment W refers to an assignment of mixing coefficients $w_{e,e'} \in F$, one for each pair of edges (e, e') with $e' \in E \cup \{s_1, \ldots, s_h\}$, $e \in E$, and head(e') = tail(e). The global coding vectors resulting from a linear network coding assignment W are the set of row vectors of the matrix $Q = (I - A)^{-1}B$, where $W = [A \ B]$.

In a linear network coding assignment W, the rank of a node v, rank$_v(W)$, refers to the rank of a linear space spanned by the global coding vectors for incoming edges of v, i.e.,

$$\text{rank}_v(W) \equiv \text{rank}\{q_e(W), \text{head}(e) = v\}. \tag{5.7}$$

For a receiver to decode the received messages successfully, the rank of the receiver must be no less than the coding dimension h. A valid linear network coding assignment is a linear network coding assignment that all the receivers can decode successfully.

Given a network topology, there may be multiple valid linear network coding assignments. We use the network topology shown in Fig. 5.1 as an example. In this example, $h = 2$ and the finite field is $F = GF(2)$. The set of equations of symbols on each edge are

$$y_{e1} = y_{s1};$$

$$y_{e2} = y_{s2};$$

$$y_{e3} = y_{e1};$$

$$y_{e4} = y_{e1};$$

$$y_{e5} = y_{e2};$$

$$y_{e6} = y_{e2};$$

$$y_{e7} = y_{e3} + y_{e5};$$

$$y_{e8} = y_{e7};$$

$$y_{e9} = y_{e7};$$

The matrix W is

$$W = [A_{9\times9} \ B_{9\times2}] = \begin{pmatrix} 0 & 0 & 0 & 0 & 0 & 0 & 0 & 0 & 0 & | & 1 & 0 \\ 0 & 0 & 0 & 0 & 0 & 0 & 0 & 0 & 0 & | & 0 & 1 \\ 1 & 0 & 0 & 0 & 0 & 0 & 0 & 0 & 0 & | & 0 & 0 \\ 1 & 0 & 0 & 0 & 0 & 0 & 0 & 0 & 0 & | & 0 & 0 \\ 0 & 1 & 0 & 0 & 0 & 0 & 0 & 0 & 0 & | & 0 & 0 \\ 0 & 1 & 0 & 0 & 0 & 0 & 0 & 0 & 0 & | & 0 & 0 \\ 0 & 0 & 1 & 0 & 1 & 0 & 0 & 0 & 0 & | & 0 & 0 \\ 0 & 0 & 0 & 0 & 0 & 0 & 1 & 0 & 0 & | & 0 & 0 \\ 0 & 0 & 0 & 0 & 0 & 0 & 1 & 0 & 0 & | & 0 & 0 \end{pmatrix}$$

The matrix $Q = (I - A)^{-1}B$ is

$$Q_{9\times2} = (I - A)^{-1}B = \begin{pmatrix} 1 & 0 & 0 & 0 & 0 & 0 & 0 & 0 & 0 \\ 0 & 1 & 0 & 0 & 0 & 0 & 0 & 0 & 0 \\ 1 & 0 & 1 & 0 & 0 & 0 & 0 & 0 & 0 \\ 1 & 0 & 0 & 1 & 0 & 0 & 0 & 0 & 0 \\ 0 & 1 & 0 & 0 & 1 & 0 & 0 & 0 & 0 \\ 0 & 1 & 0 & 0 & 0 & 1 & 0 & 0 & 0 \\ 1 & 1 & 1 & 0 & 1 & 0 & 1 & 0 & 0 \\ 1 & 1 & 1 & 0 & 1 & 0 & 1 & 1 & 0 \\ 1 & 1 & 1 & 0 & 1 & 0 & 1 & 0 & 1 \end{pmatrix} \begin{pmatrix} 1 & 0 \\ 0 & 1 \\ 0 & 0 \\ 0 & 0 \\ 0 & 0 \\ 0 & 0 \\ 0 & 0 \\ 0 & 0 \\ 0 & 0 \end{pmatrix} = \begin{pmatrix} 1 & 0 \\ 0 & 1 \\ 1 & 0 \\ 1 & 0 \\ 0 & 1 \\ 0 & 1 \\ 1 & 1 \\ 1 & 1 \\ 1 & 1 \end{pmatrix}.$$

With this linear network coding assignment, receiver $t1$ can decode the two source symbols since it observes two symbols with linearly independent global coding vectors [1, 0] and [1, 1]. Similarly, receiver $t2$ can decode the two source symbols as well.

In the above, a linear network coding assignment is defined by specifying the mixing coefficient matrix W. Following theorem gives the sufficient and necessary condition that a linear network coding assignment W can achieve the multicast capacity.

Theorem 2. *A capacity-achieving linear network coding assignment for an acyclic graph $G = (V, E)$ exists if and only if we can assign vectors $\{q_e\}_{e \in E}$ to the edges that satisfy:*

$$q_e \in \text{span}\{q_{e'}, \ \text{head}(e') = \text{tail}(e), \forall e \in E\} \tag{5.8}$$

$$\text{rank}\{q_{e'}, \ \text{head}(e') = t\} = \text{multicast\_capacity}, \forall t \in T. \tag{5.9}$$

5.1.4 Deterministic Network Coding vs. Random Network Coding

To apply linear network coding to multicast networks is to find a capacity-achieving linear network coding assignment for a multicast network. There are two ways to fulfill this task: deterministic network coding and random network coding.

Deterministic network coding adopts a centralized way to calculate the encoding mixing coefficients for each node in the network given the information about the network topology, source and receivers. Jaggi et al. proposed a polynomial deterministic linear network coding construction algorithm in [5]. After the linear network coding assignment is determined, it is distributed around the network. Each node then encodes the incoming messages based on the linear network coding assignment during the whole multicast session. As the linear network coding assignment is calculated centrally based on the complete information about the multicast session, all the receivers are guaranteed to be able to decode properly. Moreover, the required field size can be as small as the number of receivers [16]. A drawback of deterministic linear network coding is its dependence on the stableness of the system topology. Once the topology is changed, the whole linear network coding assignment needs to be recalculated.

On the contrary, random network coding uses a complete distributed way to determine the encoding mixing coefficients for each node. Each node even does not need to collect any local information. The coefficients is randomly generated for each node. The random coefficients is attached to the corresponding encoded messages. After receiving the encoded messages, the relay nodes will encode again with a set of new generated random coefficients and replace the coefficients in the message with the new ones. At the end, the receivers will try to decode the messages based on the coefficients attached in the messages. Due to the randomness of the coefficients, there is a non-zero possibility that the receiver cannot decode successfully. In this case, the receiver has to receive more messages to perform decoding. Obviously, if we increase the field size for the coefficients, the probability of failing to decode is reduced. The strength of random linear network coding is its resilience under dynamic network topology. As each node generated the coefficients independently and randomly, the topology change has no impact on coefficients generation. For details about random linear network coding, please refer to [21].

5.2 Network Coding in Peer-to-Peer File Sharing (PPFEED)

Many emerging web-based applications involve one source (server) and multiple destinations (receivers). However, due to the lack of multicast support over the Internet, these applications usually suffer from the scalability problem, which limits the number of receivers involved. By incorporating peer-to-peer technology into web-based applications, the scalability problem can be eliminated.

In this section, we consider applying peer-to-peer technology and network coding to file sharing services, in which a web server or a file server holds a file that is requested by multiple clients (receivers). The topology of such an application can be represented by a multicast network. When peer-to-peer technology is used, the receivers help forward the file for each other besides receiving the file. In [23], linear network coding was applied to ALM, in which a rudimentary mesh graph is first constructed, and on top of it a rudimentary tree is formed. Then a multicast graph is constructed, which is a subgraph of the rudimentary mesh and a supergraph of the rudimentary tree. The multicast graph constructed this way is 2-redundant, which means that each receiver has two disjoint paths to the source. By taking advantage of the 2-redundancy property of the multicast graph, a light-weight algorithm generates a sequence of 2-dimensional transformation vectors which are linearly independent. These vectors are assigned to the edges as their edge functions. However, the paper did not discuss how to process dynamic joining or leaving of peers, while dynamic membership is a common phenomenon in peer-to-peer networks. Moreover, the 2-redundancy property limits the minimum cut of the multicast graph, which in turn limits network throughput. In [12], random network coding was applied to content distribution, in which nodes encode their received messages with random coefficients. Compared to deterministic network coding, random network coding is inherently distributed. In random network coding, nodes can determine the edge functions of its outgoing edges independently by generating random coefficients for the edge functions. The advantage of random network coding is that there is no control overhead to construct and maintain a linear coding assignment among nodes. However, the global encoding vectors of a receiver's incoming edges may not be linearly independent. In other words, a receiver may not recover the original messages even it receives k or more messages (here k is the multicast capacity of the multicast network). To reduce the probability of failing to decode messages, it is required to encode over a very large field. Another drawback of random network coding is the increased data traffic. As there is no deterministic path for data delivery, all the nodes take part in relaying the data to the receivers even if it is not necessary. As a result, the same message may be transmitted through the same link multiple times.

In this section, we will discuss a new file sharing service over peer-to-peer networks by utilizing network coding. We call it *Peer-to-Peer FilE sharing based on nEtwork coDing*, or *PPFEED* for short. PPFEED utilizes a special type of network with a regular topology called *combination network*. It was demonstrated in [29] that when the network size increases, this type of network can achieve unbounded network coding gain measured by the ratio of network throughput with network coding to that without network coding. The basic idea of PPFEED is to construct an overlay network over the source and the receivers such that it can be decomposed into multiple combination networks. Compared to [23], PPFEED can accommodate dynamic membership and construct a much simpler overlay network topology in different k values. Compared to [12], the network coding construction scheme of PPFEED is deterministic, which means that the validity of the network coding assignment is guaranteed. The data traffic is then minimized so that the same messages are transmitted through an overlay link at most once. Also, system reliability is improved dramatically with little overhead. In addition, PPFEED can be extended to support link heterogeneity and topology awareness.

5.2.1 Deterministic Linear Coding over Combination Networks

One of the advantages of network coding is that it can increase network throughput by achieving multicast capacity. *Network coding gain* is defined as the ratio of throughput with network coding to that without network coding. Clearly, it is always greater than or equal to 1. When applying network coding to a multicast network, the gain depends on the topology of the multicast network. For example, if the topology of a multicast network is a tree, i.e., the source is the root of the tree and the receivers are the leaves of the tree, network coding gain is 1. Intuitively, if each receiver has roughly the same number of disjoint paths to the source, the more overlap among the paths, the larger network coding gain can be achieved.

A combination network is a multicast network with a regular topology. The topology of a combination network is a regular graph which contains three types of nodes: *source node, relay node,* and *receiver node.* A combination network contains a source node which generates messages, n relay nodes which receive messages from the source node and relay them to the receiver nodes, and C_n^k receiver nodes which receive messages from the relay nodes. There are n links connecting the source node to the n relay nodes, respectively. For every k nodes out of the n relay nodes, there are k links connecting them to a receiver node. Since there are a total of C_n^k different combinations, the number of receiver nodes is C_n^k. The capacity of each link is 1. Figure 5.2 shows a combination network for $n = 4$ and $k = 2$, usually denoted as a C_4^2 combination network. For clarity, the nodes are arranged into three layers: the first layer contains only one node, the source node; the second layer contains n relay nodes; and the third layer contains C_n^k receiver nodes.

Combination networks have good performance with respect to network coding gain. It was proved [29] that network coding gain is unbounded when both n and k approach infinity. For a C_n^k combination network, the multicast capacity is k as each

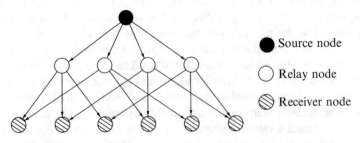

Fig. 5.2 Combination network C_4^2

receiver node has k disjoint paths to the source node. This implies that to achieve the multicast capacity in a combination network, the minimum subgraph to deploy network coding is the entire topology of the combination network.

We have discussed the good properties of combination networks. The next issue is how to find a valid linear network coding assignment on a combination network. As mentioned earlier, existing network coding methods can be classified into two categories: random method and deterministic method. Random network coding [32] assigns random mixing coefficients to the edges. This method can only obtain a valid network coding assignment with a probability less than 1. In [5], a polynomial deterministic algorithm was proposed to construct a valid linear coding assignment for any multicast network. However, it is a centralized algorithm which requires all the topology information in advance. Moreover, since it is designed for general topologies, it involves many complex processes such as keeping track of the paths between the source and the receivers and maintaining an extra array to facilitate linear independence test. On the other hand, for a multicast network with a regular topology like a combination network, it is possible to design a tailored linear network coding assignment which takes advantage of the regular topology to achieve both effectiveness and efficiency. In this section, we propose a simple deterministic linear coding construction scheme which can guarantee a valid linear network coding assignment and achieve $O(1)$ time complexity.

Before diving into details of the linear network coding construction, it is important to determine an appropriate value for k, which is the multicast capacity of a combination network. On one hand, given the number of relay nodes n, the network coding gain reaches its maximum when $k = n/2$ [29]. On the other hand, k should not be too large because a large k will increase the link stress and overhead. Therefore, as a tradeoff, in this chapter we only consider combination networks with $n = 4$, $k = 2$ or $n = 6$, $k = 3$ where the network coding gain is 1.5 and 2, respectively. The n and k determine the size of the combination network. As we will see later, the size of the overlay network is independent of the size of the combination network. If we apply the same n and k to different overlay networks of various sizes, the performance difference is very small.

As mentioned earlier, constructing a valid linear network coding assignment is equivalent to assigning each edge a global encoding vector such that the vectors of the incoming edges of a receiver are linearly independent. Suppose the source has k symbols to multicast to the receiver nodes. In a C_n^k combination network, there are a total of $n + kC_n^k$ edges. The first n edges connect the source node to n relay nodes with global encoding vectors $V_1', V_2', \ldots, V_n'$. For each relay node, there are kC_n^k/n outgoing edges connecting it to kC_n^k/n receiver nodes. As each relay node has only one incoming edge, the global encoding vector of its outgoing edge is the global encoding vector of its incoming edge multiplied by a constant. Since the constant does not change the linear independence property, we only consider the case that the constant is 1. Each receiver has k incoming edges whose global encoding vectors are k vectors out of the n vectors $V_1', V_2', \ldots, V_n'$. For a receiver node to decode the original messages, the k global encoding vectors must be linearly independent. Thus the main issue is how to find a set of n k-dimensional vectors such that every k vectors of the n vectors are linearly independent.

Suppose $GF(q)$ is a given Galois field, where $|GF(q)| = q$, $GF = \{0, 1, 2, \ldots, q-1\}$, $q \geq n$. We give the rules for the linear network coding assignment construction for C_n^2 and C_n^3 combination networks as follows:

1. The linear network coding construction scheme for C_n^2 combination network is to assign vectors $(1, \alpha_1), (1, \alpha_2), \ldots, (1, \alpha_n)$, where $\alpha_1, \alpha_2, \ldots, \alpha_n$ are different symbols in $GF(q)$, to n edges connecting to n relay nodes as global encoding vectors. The global encoding vectors of the edge outgoing from one relay node is the same as the global encoding vector of the edge incoming to the relay node.
2. The linear network coding construction scheme for C_n^3 combination network is to assign vectors $(1, \alpha_1, \alpha_1^2 \bmod q)$, $(1, \alpha_2, \alpha_2^2 \bmod q)$, $\ldots$, $(1, \alpha_n, \alpha_n^2 \bmod q)$, where $\alpha_1, \alpha_2, \ldots, \alpha_n$ are different symbols in $GF(q)$, to n edges connecting to n relay nodes as global encoding vectors. The global encoding vectors of the edge outgoing from one relay node is the same as the global encoding vector of the edge incoming to the relay node.

We have the following theorem concerning the linear network coding assignment.

Theorem 3. *The proposed linear network coding assignment for combination networks is valid.*

Proof [3]. Case 1: $k = 2$. For any two vectors $(1, \alpha)$, $(1, \beta)$, we evaluate the determinant of matrix $l \begin{pmatrix} 1 & \alpha \\ 1 & \beta \end{pmatrix}$. We have

$$\begin{vmatrix} 1 & \alpha \\ 1 & \beta \end{vmatrix} = \beta - \alpha. \tag{5.10}$$

Since $\alpha \neq \beta$, the determinant is not equal to 0. We conclude that the two vectors are linearly independent.

Case 2: $k = 3$. For any three vectors $(1, \alpha, \alpha^2 \bmod q)$, $(1, \beta, \beta^2 \bmod q)$ and $(1, \gamma, \gamma^2 \bmod q)$, where $\gamma > \beta > \alpha$, the determinant of matrix $\begin{pmatrix} 1 & \alpha & \alpha^2 \\ 1 & \beta & \beta^2 \\ 1 & \gamma & \gamma^2 \end{pmatrix}$ is

$$\begin{vmatrix} 1 & \alpha & \alpha^2 \\ 1 & \beta & \beta^2 \\ 1 & \gamma & \gamma^2 \end{vmatrix} = \beta\gamma^2 + \alpha\beta^2 + \alpha^2\gamma - \alpha^2\beta - \alpha\gamma^2 - \beta^2\gamma$$

$$= \beta\gamma(\gamma - \beta) + \alpha\beta(\beta - \alpha) + \alpha\gamma(\alpha - \beta + \beta - \gamma)$$

$$= (\beta - \alpha)(\gamma - \beta)(\gamma - \alpha). \tag{5.11}$$

Since $\alpha \neq \beta \neq \gamma$, the determinant is not equal to 0. We conclude that the three vectors are linearly independent. ∎

---

[3]Since the mod operation is distributive over addition and multiplication, we omit mod$q$ in the equations.

5.2.2 Peer-to-Peer File Sharing Based on Network Coding (PPFEED)

Suppose there is a server holding the file to be distributed. Peers interested in the file form an overlay network through which the file is distributed. The construction of the overlay network is based on the idea of combination networks. Similar to the combination networks, there are two system parameters n and k in the overlay network. The server encodes the file into n different messages using the linear network coding assignment given in the previous section. Thus any k messages out of the n messages can be used to decode the original file. The peers are divided into n disjoint groups and each group is assigned a unique *group ID*. Each group of peers is responsible for relaying one of the n encoded messages. To accelerate the distribution of the messages, peers in the same group are connected by an unstructured overlay network according to some loose rules. Each peer in a group is connected to at least other $k-1$ peers which are in $k-1$ different groups, respectively.

Although the topology is constructed based on combination networks, it is more a mesh network than a multi-tree network. If we color peers with n different colors, the entire overlay topology can be looked as a mesh with the constraint that each peer must be connected to at least k peers with k different colors. Therefore, PPFEED enjoys the scalability and resilience of a mesh network. In Sect. 5.2.9, we will quantify its performance through simulations.

Figure 5.3 shows an example of the overlay network constructed by PPFEED for $n = 3$ and $k = 2$. There are three groups of peers which are colored in black, blue, and red, respectively. The colored links represent corresponding messages with arrows pointing to the transmission directions. We can see that each peer has at least two links with different colors pointed to itself. All the peers are able to decode the received messages.

The overlay links are added to the system on demand as new peers join the system. There are two key issues in the overlay network construction: first, how to connect the peers in the same group; second, how to connect the peers in different groups.

Fig. 5.3 An example overlay network constructed by PPFEED

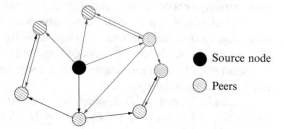

Source node

Peers

The first issue involves how to distribute the n messages in a scalable way. When the number of peers is small, it is acceptable to connect all the peers to the source node, i.e., the server. When the number of peers increases, the peers resort to each other instead of the server to distribute the messages. Here we adopt some loose rules similar to the unstructured peer-to-peer network Gnutella [10]: for a newly joined peer, it is responded by a random list of the existing peers, and the new peer creates connections with these peers. The reason is three folds. First, it is resilient to peer join. A new peer can be connected to any existing peer. Second, the overlay construction and maintenance overhead is low. There are no constraints on the overlay topology as long as it is connected. If some peers leave the system, it is convenient for the rest of the peers to reconnect through the server. Third, it is easy to flood messages on the resulting overlay network.

The second issue involves how the peers receive other $k - 1$ messages to decode the messages. As any k different messages can be used to decode, we only need to find $k - 1$ peers in different groups. Since there are $n - 1$ eligible groups, we first connect the peer to $k - 1$ random peers in different groups. Then perform a local topology adjustment to find the $k - 1$ peers such that the average latency between the peer and $k - 1$ peers is minimized.

Some erasure codes, such as Reed-Solomon or Tornado, have similar property to PPFEED in terms of decoding the original messages from a group of encoded messages. In [26], the authors have applied Tornado code to reliable distribution of bulk data to a variety of heterogeneous population of receivers. The erasure codes encode the original messages into a sequence of encoded messages such that any subset of the encoded messages can be used to reconstruct the original messages as long as the size of the subset is sufficiently large. However, they have different objectives and applications. The erasure codes are designed to avoid retransmissions, while network coding is used to improve system throughput. In some applications, retransmissions may be impossible or undesirable. For example, in multicast applications, retransmission requests may greatly overwhelm the sender if the group size is large. In satellite networks, where the back channel has high latency and limited capacity, retransmission requests are costly and unreliable. With the erasure codes, the sender can keep sending the sequence of encoded messages regardless of whether the receivers receive them or not. As long as the receivers receive enough encoded messages, they can extract the original messages from the received messages. On the other hand, network coding is applied when multiple flows share a common link where the messages from different flows can be mixed together to save bandwidth. Network coding can be applied to multicast networks or wireless networks where messages sharing common links can be encoded together to improve system throughput.

As mentioned earlier, we consider the combination networks with $n = 4, k = 2$ or $n = 6, k = 3$ in this chapter. When we apply PPFEED to a large overlay network, as n and k are fixed, the number of peers in the same group increases. Each peer is still connected to the other $k-1$ peers in different groups. The scalability of PPFEED is derived from the scalability of the overlay network composed of peers within the same group. As described earlier, the overlay network is constructed by some loose rules to accommodate scalability. Therefore, PPFEED can be applied to overlay networks of various network sizes without performance degradation.

5.2.3   Peer Joining

Suppose the server is well-known whose IP address is known to all the peers by some address translation service such as DNS. When a peer wants to retrieve a file hosted by the server, it initiates a join process by sending a JOIN request to the server.

The server keeps track of the number of peers in each group and maintains a partial list of existing peers in the group and their IP addresses. The purpose of maintaining a partial list instead of a full list is to achieve a balance between scalability and efficiency. On one hand, when the network size is large, it is resource-consuming to maintain a full list of peers in the server. On the other hand, if the list is too short, it may cause much longer latency for new peers to join when the peers in the list crash. Although the server is responsible for bootstrapping the peers, it will not be the bottleneck of the system because once each peer receives the list, it communicates with other peers for topology construction and data dissemination. When the server receives a peer's join request, it assigns the peer to a group such that the numbers of peers in different groups is balanced. Then the server sends the list of peers of that group to the joining peer and updates the number of peers in that group.

After receiving the list of peers, the new peer will contact them and create overlay links with them. These peers are called intra-neighbors of the new peer because they are within the same group. In contrast, the neighbors which are in different groups are called inter-neighbors. The new peer asks one of its intra-neighbors picked randomly to provide a list of its inter-neighbors. The new peer then takes the list of peers as its inter-neighbors.

The topology of the peer-to-peer network can be considered as a combination of multiple unstructured peer-to-peer networks, each of which is composed of the peers within the same group. The topology within one group is arbitrary as long as it is connected. The only constraint is on the edges between different groups. It is required that each peer is connected to at least $k-1$ peers in $k-1$ different groups, respectively.

5.2.4   Local Topology Adjustment

As the overlay topology is formed by always connecting the new peers to a random list of existing peers in the network, the overlay links among peers may not be good with respect to latency or link stress. To alleviate the performance degradation, we optimize the overlay topology by a process called *local topology adjustment*.

The idea of local topology adjustment is to replace the direct neighbors with peers with better performance through a local search. Each peer periodically sends QUERY messages to its neighbors. There are two types of QUERY messages: one for intra-neighbor searching and one for inter-neighbor searching. Suppose peer u sends out a QUERY message to one of its neighbors, say, v. After receiving the QUERY message, node v will send a RESPONSE message back to u and a QUERY-2 message to each of its neighbors except for u. The RESPONSE message includes a TIMESTAMP field which records the time when node v sends the RESPONSE message. After node u receives the message, it calculates the latency between u and v by the difference between the time when the RESPONSE message is received and the time indicated by the TIMESTAMP in the RESPONSE message. The QUERY-2 message includes the IP address of node u. After a neighbor of v, say, w, receives the QUERY-2 message, it sends a RESPONSE-2 message to node u. It also includes a TIMESTAMP field which records the time when w sends the RESPONSE-2 message. After node u receives the message, it calculates the latency between u and w by the difference between the time when the RESPONSE-2 message is received and the time indicated by the TIMESTAMP in the RESPONSE-2 message. If the latency between u and w is less than that between u and v, node w will replace node v and become the neighbor of node u.

Figure 5.4 illustrates an example for local topology adjustment. In the figure, an overlay topology with $n + 2$ nodes connected as a list. We use the distance between nodes to represent the latency roughly. At the beginning, node v_1 is the neighbor of node u. After n steps of local topology adjustment, node w becomes the neighbor of node u as the topology on the right shows.

Local topology adjustment is a distributed light-weight overlay topology optimization process. It can detect a better neighbor after searching the nodes nearby with a search radius of 2. The process is done periodically as the topology evolves. However, in some cases, local topology adjustment cannot find the best neighbor with the shortest latency. Figure 5.5 shows one of these cases, where although node u and node w are close to each other, they cannot detect each other through local topology adjustment. In Sect. 5.2.8.2, we describe a landmark-based solution to solve this topology mismatch problem systematically.

Fig. 5.4 Illustration of local topology adjustment

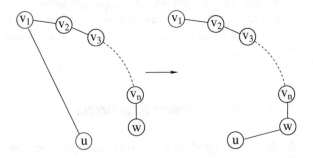

Fig. 5.5 An example that
local topology adjustment
does not work

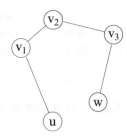

5.2.5   Peer Leaving

There are two types of peer leaving: friendly or abruptly. Friendly leaving means
that the leaving peer initiates a leaving process so that the system is aware of its
leaving and can make necessary updates accordingly. Abruptly leaving means that
the leaving peer leaves the system without any notification, mainly due to link crash
or computer crash.

For the friendly leaving, the leaving peer will initiate a leaving process by sending
LEAVE messages to both of its intra-neighbors and inter-neighbors. The leaving of
the peer may impair the connectivity between peers in the same group. To rebuild
the connectivity, the intra-neighbors will initiate the join process with the group
ID in the join request. After receiving a join request with a designated group ID,
the server temporarily acts as one of the intra-neighbors of the peer to guarantee
the connectivity. The topology is further adjusted by the local topology adjustment
process. Here the server acts as a connection "hub" for the peers that were connected
to the leaving peer. This may increase the data forwarding burden on the server
temporarily. Nevertheless, the situation will be improved after the local topology
adjustment process is completed. Moreover, it can achieve strong robustness with
little control overhead. For example, it can handle concurrent peer leavings. The
connectivity is rebuilt after the server receives all the join requests from the intra-
neighbors of the leaving peers.

After receiving the LEAVE message from the leaving peer, the inter-neighbors
will ask one of its intra-neighbors for a new inter-neighbor to replace the leaving
one. In addition, the leaving peer will also send a LEAVE message to the server.
This LEAVE message will make the server update the number of peers in the group.

For the abruptly leaving, peers send HELLO messages to its neighbors periodically and maintain a HELLO timer for each neighbor. Receiving a HELLO
message triggers a reset of the corresponding HELLO timer. The neighbors detect
the abruptly leaving by the timeout of the HELLO timer. Similarly, intra-neighbors
initiate the join process after detecting the sudden leave. Inter-neighbors ask their
intra-neighbors for a replacement of the left peer. Moreover, one of the neighbors is
chosen to send a LEAVE message to the server on behalf of the abruptly leaving peer
so that the server can update the number of the peers in the group. To minimize the
selection overhead, we choose the inter-neighbor whose group ID is next to that of
the leaving peer to perform this task. For example, if we assume the encoding vector

for the group corresponding to the leaving peer is $(1, \alpha)$ (or $(1, \alpha, \alpha^2 \ mod \ q)$), inter-neighbor corresponding to the group whose encoding vector is $(1, (\alpha+1) \ mod \ n)$, or $(1, (\alpha+1) \ mod \ n, ((\alpha+1) \ mod \ n)^2 \ mod \ q)$ is chosen.

5.2.6 Data Dissemination

Before sending out the file, the server needs to encode the file. The encoding is over a Galois field $GF(q)$. The file is divided into multiple blocks with each block represented by a symbol in $GF(q)$. The first k blocks are encoded and then the second k blocks, and so on. In the case that there are not enough blocks to encode, the file is padded with zero string. Assuming the field size is $q(> n)$, each k blocks are encoded into n different messages using the linear network coding assignment given in Sect. 5.2.1. Therefore, any k messages of these n messages can be used to decode the original k blocks.

After encoding, the server sends the encoded n messages to the peers in the n groups, respectively. The group ID and the encoding function form a one-to-one mapping. Peers can learn which messages they receive based on the sender of the messages: if the sender is the server, the messages correspond to the group ID of the peer itself; if the sender is a peer, the messages correspond to the group ID of the sender.

The peers forward all the messages they receive based on the following rules:

Rule 1. If the message comes from the server, the peer forwards it to all its intra-neighbors and inter-neighbors.
Rule 2. If the message comes from one of its intra-neighbors, the peer forwards it to other intra-neighbors except for the sender.
Rule 3. If the message comes from one of its inter-neighbors, the peer does nothing.

Peers forward messages in a push style, which means that the messages are forwarded under the three rules as soon as they arrive at a peer. A peer decodes the messages right after it receives k different messages. Thus data dissemination is fast and simple in PPFEED.

5.2.7 Improving Reliability and Resilience to Churn

Besides the throughput improvement, PPFEED can provide high reliability and high resilience to *churn* which refers to frequent peer joins or leaves. All we need to do is to add a redundant overlay link for each peer.

In the previous subsections, each peer is connected to $k - 1$ inter-neighbors. The $k - 1$ messages received from them plus the message received from the source or the intra-neighbors should be sufficient to decode the original file blocks.

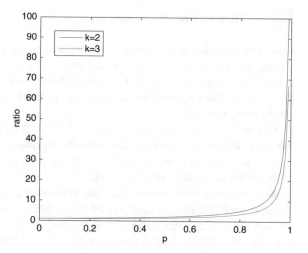

Fig. 5.6 Failure probability ratio of the old scheme to the improved scheme

However, no links are 100 % reliable. In case that the messages are lost or damaged, the sender has to retransmit the messages until they are received correctly. Clearly, retransmission will cause longer latency, larger buffer size, and reduce system throughput. PPFEED can reduce the retransmission probability by introducing a redundant link to each inter-neighbor. Now each peer is connected to k instead of $k - 1$ peers in different groups. If one of the links fails, the peer can still decode the original file blocks based on the remaining $k - 1$ messages. The failure probability of this scheme can be quantitatively analyzed as follows. Suppose each overlay link will fail with probability $1 - p$. In the old scheme, the peer fails to decode with probability $P_{\text{failure}} = 1 - p^{k-1}$, while in the improved scheme, the peer fails to decode with probability $P'_{\text{failure}} = 1 - (p^k + k(1 - p)p^{k-1})$. The ratio of the two probabilities is $P_{\text{failure}}/P'_{\text{failure}} = (1 - p^{k-1})/(1 - (p^k + k(1 - p)p^{k-1}))$. Figure 5.6 plots the curves of the ratio for different p values when $k = 2$ and $k = 3$. We can see that when $p = 0.9$, the failure probability of the old scheme is about ten times of the improved scheme.

Another advantage of connecting peers with more inter-neighbors than necessary is that it can improve the system resilience to churn. The rationale is similar to the reliability improvement. For example, if we connect a peer with k other peers, it can tolerate one peer's leave without affecting the download speed. Peers experiencing unstable neighborhood should be connected to more than one redundant inter-neighbors to alleviate the performance degradation.

5.2.8 Some Extensions

In this section, we extend PPFEED to support some extra features, such as link heterogeneity and topology awareness.

5.2.8.1 Support Link Heterogeneity

Link heterogeneity is a common phenomenon in networks. In peer-to-peer networks, link heterogeneity refers to the fact that peers have different access link capability. Common access links include dial-up connections, ADSL, and high speed cable. The ratio of the link capacities of the fastest link to the slowest link can be more than 1000.

Due to the link capacity imbalance, if a peer with high link capacity is receiving messages from a peer with low link capacity, its download speed is upper bounded by the download speed of the low link capacity peer. In other words, the bandwidth of the high link capacity peer is wasted.

Now consider how to maximize the utilization of link capacity of each peer. Suppose the bottleneck is always on the access links. Thus we can construct as many as possible overlay links connected to the peer as long as its access link capacity permits. As mentioned earlier, the overlay construction between the peers in the same group is arbitrary. We take advantage of this property to construct an overlay network such that the number of links of a peer is proportional to its link capacity. For example, if a peer has much higher link capacity than other peers that have low link capacities, the topology of the overlay network is a star with the high link capacity peer in the center.

In practice, every peer first reserves link capacity for the k links that are used to connect inter-neighbors. The list kept in the server includes the peers with higher link capacities. Peers periodically send messages to the server that include their remaining link capacities for the server to update the list.

5.2.8.2 Support Topology Awareness

Since overlay networks are logical networks on top of physical networks, the overlay links are logical links. Each logic link is composed of one or more physical links. The overlay links are added arbitrarily as needed. As a result, the topology of the overlay network may be different from the topology of the physical network. Two nodes which are close to each other in the overlay network may be far away in the physical network. Such topology mismatch may greatly increase the *link stress* and degrade the performance. Here link stress is defined as the number of copies of a message transmitted over a certain physical link.

In Sect. 5.2.4, we mentioned that peers can use local topology adjustment to find peers with better performance than current neighbors. If we define the performance to be the latency, then peers can dynamically adjust the local overlay topology to alleviate the mismatch. However, a problem with this method is that its convergence speed is slow and its accuracy is limited. We need a more efficient and accurate method to minimize the mismatch. Here we propose a *topology clustering scheme* by adopting the idea of the binning scheme introduced in [18] to construct a topology aware overlay network.

In this scheme, the server is responsible for choosing some peers as *landmarks*. Each new peer will receive the list of landmarks before it receives the list of peers. The new peer sends probe messages to the landmarks to learn the distances between them and itself. The landmark peers are listed in an ascending order of distances. The ordered list acts as the coordinate of the peer in the system. The coordinate is sent to the server, then the server assigns the new peer a group ID based on its coordinate. Peers with the same coordinates form a cluster. The heuristic rules of assigning peers to groups are: each cluster has at least k different groups; the peers in the same group should span as few clusters as possible. The first rule is to guarantee that peers can receive enough messages to decode within its cluster. The second rule is to minimize the number of links across clusters. To implement this, the server should keep track of the numbers of different groups in each cluster. After receiving a peer's join request and its coordinate, the server first checks whether the corresponding cluster has k different groups. If yes, the peer is assigned to one of the groups such that the numbers of peers of different groups is as balanced as possible. Otherwise, the peer is assigned to a group which is different from the existing groups in the cluster.

In addition, every two landmark peers should not be too close to each other. A new peer cannot be a landmark if its coordinate is the same as one of the existing landmark peers. A landmark peer is removed from the landmark peers if it has the same coordinate as another landmark peer.

5.2.9 Performance Evaluations

In this section, we study the performance of PPFEED through simulations. We compare PPFEED with a peer-to-peer multicast system called Narada [31] and a peer-to-peer file sharing system called Avalanche [12]. Narada first constructs an overlay mesh spanning over all the peers. The overlay mesh is a richer connected graph which satisfies some desirable performance properties. The multicast tree is a spanning tree on top of the mesh and is constructed on demand of the source peer. Avalanche is a peer-to-peer file sharing system based on random network coding. We choose these two schemes as the comparison counterparts in order to evaluate the benefit a tailored deterministic network coding brings.

The simulation adopts following three performance metrics:

Throughput: throughput is defined as the service the system provides in one time unit. Here we let different systems transmit the same file, thus throughput can be simply represented by the time consumed by the transmission. The shorter the time consumed, the higher the throughput. We start transmitting the file from time 0. Then the consumed time is the time when the peers finish receiving the file, denoted by *finish time*.

Reliability: this performance metric is used to evaluate the ability of the system to handle errors. We use the *number of retransmissions* to characterize this ability. A system with higher reliability will have a smaller number of retransmissions, and thus higher throughput.

Link stress: link stress is defined as the number of copies of the same message transmitted through the same link. It is a performance metric that only applies to an overlay network due to the mismatch between the overlay network and the physical network. We use it to evaluate the effectiveness of the topology awareness improvement and the efficiency of the system.

We study the performance of the system in four different configurations:

(i) Baseline configuration. In this configuration, peers have uniform link capacities, and overlay links are constructed randomly. The file is sent after the overlay network is formed.
(ii) Dynamic peer join/leave configuration. In this configuration, peers have uniform link capacities, and overlay links are constructed randomly. Peers join the system during the file transmission and stay in or leave the system after downloading the whole file.
(iii) Heterogeneity configuration. In this configuration, peers have heterogeneous link capacities, and overlay links are constructed by taking into consideration of link heterogeneity. The file is sent after the overlay network is formed.
(iv) Topology awareness configuration. In this configuration, peers have uniform link capacities, and overlay links are constructed by taking into consideration of topology mismatch. The file is sent after the overlay network is formed.

The network topologies are random graphs generated by GT-ITM [11]. We conducted the same simulations for $k = 2$ and $k = 3$. In the rest of this section, we will omit the result for $k = 2$ when it is similar to that of $k = 3$.

Baseline Configuration. In the baseline configuration, peers have uniform link capacities. The simulation is divided into two steps. First, overlay construction period: A number of random nodes are picked to join the system in sequence. Second, file transmission period: The server sends the file to the overlay network.

We plot the finish time curves of PPFEED and Narada in Fig. 5.7a. We simulate PPFEED with different k values as shown in the figure. The finish time of nodes is sorted in an ascending order. It can be seen that the average finish time of PPFEED is 15–20 % shorter than that of Narada and 8–10 % shorter than Avalanche. We notice that the finish times of Narada have a larger variance than that of Avalanche and PPFEED. This is because that in Narada, the two peers with the biggest difference in finish time are a child of the root and a leaf, respectively. This difference may be very large depending on the overlay topology. On the contrary, Avalanche and PPFEED construct a mesh to distribute the file. As a result, the distance between the highest level peer and the lowest level peer is shortened. From the figure we can see that the throughput of PPFEED is higher for $k = 3$ than that for $k = 2$. This can be explained by the fact that when $k = 3$, each peer is connected to more peers. Thus the download capacity of peers can be better utilized.

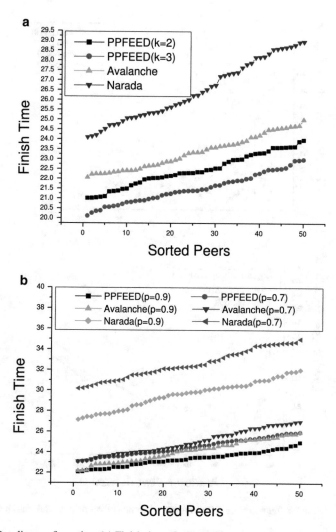

Fig. 5.7 Baseline configuration. (**a**) Finish time; (**b**) Finish time with link failures

Figure 5.7b shows the finish time when we set the physical link failure probability to $1 - p$. Note that the link failure probabilities of different physical links are independent. In the analysis in Sect. 5.2.7, we simply assumed that the link failure probability of an overlay link is $1 - p$ to simplify the analysis. However, the link failure probabilities of overlay links are dependent due to sharing common physical links. Here we use link failure probabilities of physical links to simulate real networks more accurately. We can see that the finish times of PPFEED and Avalanche are much shorter than that of Narada. With the same link failure probability, the finish time of PPFEED is slightly shorter than that of Avalanche. Compared to the previous simulation result without link failure probabilities, the

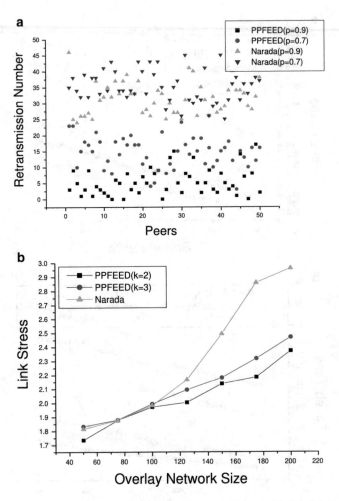

Fig. 5.8 Baseline configuration. (**a**) The number of retransmissions with link failures; (**b**) Link stress

increase of finish time is the least for Avalanche and the largest for Narada. The reduction of throughput is little for both Avalanche and PPFEED when physical links are unstable.

The number of retransmissions is shown in Fig. 5.8a. As Avalanche does not require retransmission, we only plot the curves for PPFEED and Narada. We use colored dots to denote the numbers of retransmissions of peers. We can see that Narada needs more retransmissions than PPFEED. The less the p is, the more retransmissions are needed. The average number of retransmissions of PPFEED is about 5 when $p = 0.9$ while that of Narada is about 30. Both Figs. 5.7b and 5.8a reveal that PPFEED has a good fault tolerance ability.

Each packet in Avalanche is unique, the link stress of Avalanche is always 1. We compare the link stresses of PPFEED and Narada as shown in Fig. 5.8b. When the number of peers is small, PPFEED and Narada have similar link stress. However, when the number of peers is beyond 100, the links stress of PPFEED is less than Narada. The reason for this is similar to that for the larger finish time variance of Narada. If we track a message in Narada, the paths it travels through form a tree spanning all the peers. If we track a message in PPFEED, the paths it travels through form a mesh spanning a portion of the peers (roughly peers if the overlay network is well balanced, where $1/n$ is for the peers within the same group, $(k - 1)/n$ is for the peers in different groups). Fewer overlay links will reduce the probability that the same message travels through the same physical link. The link stress of PPFEED is higher when $k = 3$ than $k = 2$. When k is larger, the number of peers in the same groups is reduced. On the other hand, each peer is connected to more peers in different groups. As a result, the increased links between different groups outnumber the reduced links due to the reduced group size. Thus the link stress is increased.

Dynamic Peer Join/Leave Configuration. In this configuration, we allow peers to join and leave the system during the file transmission to evaluate the system resilience to churn. The simulations are conducted in two scenarios: First, we let peers join the system randomly and the file transmission starts as long as there are peers requesting it. After receiving the file, peers stay in the system; Second, the peers still join the system randomly. When a peer successfully receives the file, it leaves the system right away. In both scenarios, we calculate the finish time of a peer by the difference between the time it finishes downloading the file and the time it joins the system.

Figure 5.9a shows the finish time comparison when peers stay in the system after receiving the file. We can see that the average finish time of all three schemes increases compared to the baseline configuration. For PPFEED and Avalanche, this is due to the lack of peers which help forwarding the file. For Narada, this is due to the join latency of peers. The increase of average finish time for Narada is more than that of PPFEED or Avalanche, which indicates that peer-to-peer systems achieve better resilience to dynamic joins than tree-based approaches. The peers with different finish times are not evenly distributed as in the baseline configuration. The number of peers with larger finish time increases. The largest finish time for Avalanche is close to that of PPFEED. This is because PPFEED needs to download the file from k peers from k different groups while Avalanche has no such requirements.

Figure 5.9b shows the finish time comparison when peers leave the system after receiving the file. We can see that the average finish times of PPFEED and Avalanche are increased by around 15 % while that of Narada is increased by around 50 %. Tree-based approaches are extremely vulnerable to churn as each departure will disconnect all the downstream peers and the tree needs to be rebuilt. Both PPFEED and Avalanche have similar resilience under the churn. However, PPFEED achieves slightly higher throughput. It indicated PPFEED achieves great resilience

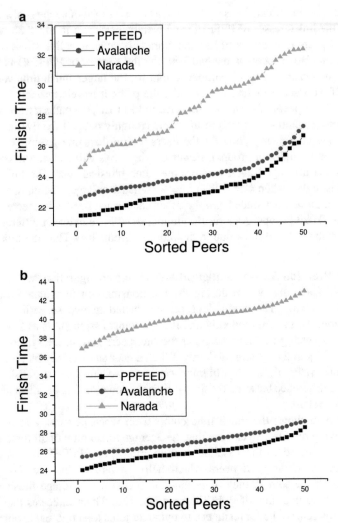

Fig. 5.9 Finish time of the dynamic peer join/leave configuration. (**a**) Peers stay in the system after receiving the file; (**b**) Peers leave the system after receiving the file

under dynamic peers join/leave. Although PPFEED adopts deterministic network coding, the overlay topology in PPFEED is quite flexible. In addition, by adding redundant links, the resilience can be improved dramatically.

Heterogeneity Configuration. Now we evaluate the ability of PPFEED to handle peers with heterogeneous link capacities. The random topologies generated by GT-ITM are flat random graphs with high speed links. Peers are added to the nodes randomly with each peer connected to one node by an access link and the access link is set to a certain link capacity. We set the peers with heterogeneous link capacities

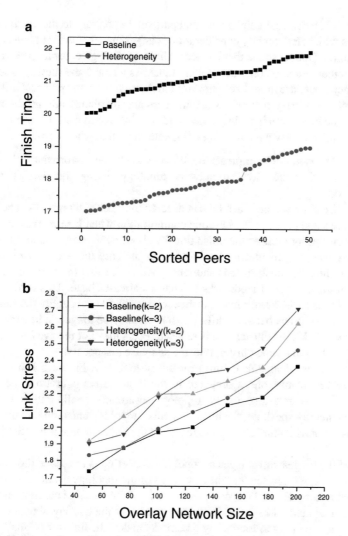

Fig. 5.10 Heterogeneity configuration. (**a**) Finish time; (**b**) Link stress

such that 1/3 with the highest link capacities, 1/3 with the lowest link capacities, and 1/3 with the medium link capacities. The highest link capacity is ten times of the lowest one.

Figure 5.10a shows the finish time comparison of the heterogeneity configuration between the overlay construction with heterogeneity consideration and without heterogeneity consideration. We can see that the finish time with heterogeneity consideration is much shorter than the baseline which does not consider heterogeneity. However, the variance of the finish time is almost the same. It indicates that the peers with higher link capacities are helpful to increase system throughput, but they cannot reduce the finish times of themselves.

Figure 5.10b shows the link stress comparison. In contrary to the finish time, the link stress with heterogeneity consideration is larger than that without heterogeneity consideration. One reason is that the access links of the high capacity peers are used by many other peers to construct overlay links. As a result, the messages sent by a high capacity peer are more likely transmitted through the same physical link. When the size of the overlay network is small, peers are distributed around the network sparsely. The link stress is mainly determined by the positions of the peers. In some cases, the link stress when $k = 2$ is even greater than that when $k = 3$.

Topology Awareness Configuration. We now study the performance of PPFEED when considering the physical network topology during the overlay network construction.

Figure 5.11a shows the finish time at different number of landmarks. The highest curve is the same curve in the baseline configuration when $k = 3$. We can see that topology clustering reduces the finish time by about 10 % compared to that without topology clustering. Increasing landmarks can increase the accuracy of topology clustering, thus the finish time is shortened. We notice that the curve of the finish time when the number of landmarks is 8 has a staircase shape. This is because that the peers in the same cluster may finish receiving the file at roughly the same time. While the finish times between different clusters may be longer. When the number of landmarks is 12, the cluster size is reduced. Peers may not receive $k - 1$ different messages within the same cluster, thus the staircase disappears.

We set the physical link failure probability to $1 - p$. Figure 5.11b shows the number of retransmissions. Compared to the baseline configuration, the average number of retransmissions of the topology awareness configuration is slightly smaller. Generally speaking, the improvement of topology clustering on the number of retransmissions is small. The main reason of the improvement is due to redundant links.

One of the biggest advantages of topology clustering is to reduce the link stress. From Fig. 5.12, we can see that the link stress of the topology awareness configuration is reduced by about 37 % compared to that of the baseline configuration when the number of landmarks is 12. With the increase of the overlay network size, the difference is even bigger. Increasing landmarks makes the link stress smaller.

5.3   Network Coding in Heterogeneous Peer-to-Peer Streaming

Live media steaming systems are important Internet applications and contribute a significant amount of today's Internet traffic. Like bulk data distribution systems, live media streaming systems usually involve a server which hosts the media content and all the clients request the media content from the server. However, there is a fundamental difference between them. Live media streaming systems

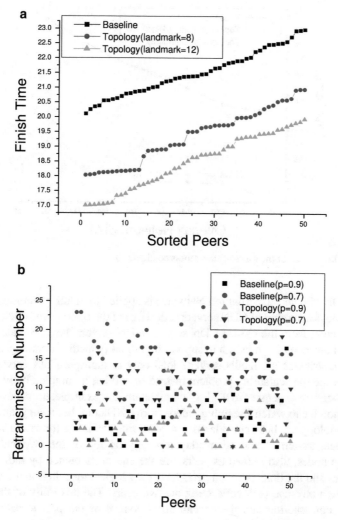

Fig. 5.11 Topology awareness configuration. (**a**) Finish time; (**b**) The number of retransmissions

require real-time data delivery and can tolerate data loss to some extent. While bulk data distribution systems require reliable data delivery and can tolerate delay to some extent. As we will see later, this requirement difference leads to different consideration in system design.

A naive way to implement such systems is to create a unicast connection between each client and the server. However, this approach scales poorly because a surge of client population could easily overwhelm the media server's resources or bandwidth. Network layer multicast provides an efficient way for one-to-many communication but its limited deployment on the Internet makes it impractical. A new technology

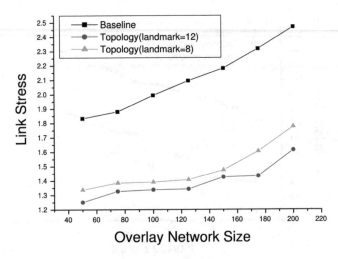

Fig. 5.12 Link stress of the topology awareness configuration

called CDN (Content Distribution Network) is applied to media streaming. Usually CDN deploys a number of CDN servers at the edge of the Internet and clients request media content from the closest CDN servers. CDN servers have dedicated storage space and out-bound bandwidth to support high-quality media streaming. However, as CDN distributes the media server's load only to multiple CDN servers, it can only alleviate the scalability problem instead of solving it completely. In addition, CDN servers are expensive to deploy and maintain. For example, users have to pay a subscription fee to watch streaming videos from CNN.com. In recent years, peer-to-peer technology has been considered as a promising candidate for media streaming. Peer-to-peer systems build an overlay topology on top of the physical network where the nodes, also known as peers, are the end hosts owned by individuals or companies and the links between peers have only logical meanings and are realized by finding a physical path connecting the two peers. The flexibility of the overlay topology construction and the decentralized control of the peer-to-peer network make it suitable for distributed applications.

When applied to media streaming systems, peer-to-peer technology can completely eliminate the scalability problem caused by the server–client transmission model. Peer-to-peer media streaming systems employ the clients (peers) to help forward the media content, that is, the systems leverage the upload bandwidth of peers to distribute the media content. Peers forward the media content after they receive it from the server or other peers. As most peers are individual computers which are connected to the Internet through access links with limited bandwidth, a practical peer-to-peer streaming system should meet the following requirements:

1. Accommodation of link heterogeneity. Most peers are individual computers connected to the Internet through access links. Due to different access link technologies (ADSL, Ethernet, Wireless LAN, etc), peers have diverse upload

and download bandwidths. The overlay topology should utilize the upload bandwidths in an efficient way to meet different requirements of playback quality by the peers.
2. Low end-to-end delay. Media streaming applications are real-time applications which are sensitive to end-to-end delay. The overlay topology should allow peers to receive the media content within a limited latency after the server sends it.
3. Adaptation to dynamic peers join/leave. Peers can join or leave the media streaming system at will. As the join order is random, the overlay topology may become far from the optimal one. The overlay topology should adapt itself to the changing peers in the system.

There has been much work on peer-to-peer streaming systems in the literature in recent years, see, for example, [2, 4, 13, 17, 24, 27, 38]. Based on the overlay topology construction, these peer-to-peer streaming systems can be categorized into two types: tree-based and mesh-based. Tree-based peer-to-peer streaming systems build one or more trees rooted at the server. The media content flows from the server to the peers along the trees. Multiple trees are employed to make use of the uplink bandwidth of the leave peers. Mesh-based peer-to-peer streaming systems build a loosely connected mesh where each peer establishes connection with several neighbors. Peers exchange the availability information of the media content with their neighbors and help each other to deliver the media content. Most of the works take little or no consideration of huge diversity of the link bandwidth of peers.

With the rapid expansion of the Internet, more and more individual computers are connected to the Internet which become potential peers for peer-to-peer streaming systems. Diverse and limited peer access link bandwidth is an important characteristics when these individual computers join a peer-to-peer streaming system. First, as different peers have different download bandwidths, it is desirable that peers with higher download bandwidth receive the media content at a higher bit rate while peers with lower download bandwidth receive the media content at a lower bit rate. This is achievable through the media streaming encoding technology (e.g., MDC [15] and layered encoding [6, 35]). Besides, the overlay topology should guarantee a path from the source to the peer with sufficient bandwidth. Second, the limited uplink bandwidth should be utilized wisely so that the total downloading rate is maximized.

In this section, we focus on the heterogeneity problem in peer-to-peer media streaming system. The issue is addressed in two steps. First, a topology construction scheme is proposed to optimize the overlay topology construction for peer-to-peer streaming systems with heterogeneous downloading requirements. Although the scheme is designed for live peer-to-peer media streaming systems, the result can also be applied to peer-to-peer content delivery or file downloading systems. Second, network coding is applied to heterogeneous peer-to-peer media streaming systems based on the proposed topology construction scheme.

5.3.1 Optimal Overlay Topology Construction
for Heterogeneous Peer-to-Peer Streaming Systems

The overlay topology for peer-to-peer streaming systems can be divided into two types: tree-based and mesh-based [45]. Tree-based peer-to-peer streaming systems build one or more trees for the media content delivery transmission. The direction and the transmission rate of the media content are fixed by the trees. Mesh-based peer-to-peer streaming systems build a loosely connected mesh in a sense that peers can switch from one neighbor to another any time as they want. The direction and the transmission rate of the media content delivery depend on the availability and current residual bandwidth. One thing needs to be clarified here is that a scheme constructing a mesh overlay topology is still a tree-based scheme if the mesh is formed by a union of multiple trees. In other words, the division of whether a scheme is tree-based or mesh-based is dependent on the media content transmission paths not on the resulting topology. Overall, tree-based approaches provide more stable transmission paths while mesh-based approaches provide a more flexible topology. For a heterogeneous peer-to-peer streaming system with limited uplink bandwidth, we believe the tree-based approach is more suitable than the mesh-based approach. Therefore, we will mainly discuss several typical tree-based peer-to-peer streaming systems and introduce one mesh-based peer-to-peer streaming system for completeness in this section.

DONet (Data-driven Overlay Network for live media streaming) [24] is a mesh-based data-centric peer-to-peer media streaming system in which the media content is transmitted on-demand. In DONet, peers are connected by randomly selected links among them and adjacent peers are called partners. Each peer maintains a list of partners and a buffer map which indicates the media content the peer contains. Peers continuously exchange the buffer map with their partners. The key operation in DONet is that every peer periodically exchanges buffer map with its partners and retrieves the missing media content from its partners. A heuristic scheduling algorithm is designed to determine the order and the supply partners of the media content retrieval.

Next, we introduce two tree-based approaches which arrange the peers into a hierarchical structure that implicitly defines the topology of the tree.

Developed by University of Maryland, NICE [2] is a project of designing a cooperative framework for scalably implementing distributed applications including peer-to-peer streaming systems over the Internet. The NICE protocol organizes peers into a cluster hierarchy where each layer is composed of one or more clusters of peers. For each cluster, a cluster leader is selected among the cluster members. Each layer is composed of the peers which are the cluster leaders in the lower layer. The size of a cluster is between k and $3k - 1$, where k is a constant and set to 3 in the paper. The peers in a cluster are close to each other and the peer at the cluster center is chosen as the cluster leader. The bottom layer contains all the peers while the top layer contains only one peer. The cluster hierarchy implicitly defines the media content delivery paths. When designing a peer-to-peer streaming system over NICE, the server is the top peer. A multicast tree is built for media content distribution in such a way that the root is the server and the parent of a peer is its cluster leader.

One problem in the NICE protocol is that the cluster leader is responsible for forwarding the media content to all the clusters it belongs to, so that in the worst case, the cluster leader in the top layer forwards the media content to $O(k \log_k N)$ other peers, where N is the number of peers in the system. ZIGZAG [17] is another hierarchical peer-to-peer system which limits the out-degree of a peer by an upper bound. ZIGZAG organizes the peers into a multi-layer hierarchical cluster which is similar to the NICE protocol. The difference is that ZIGZAG introduces the concept of *associate head* in the cluster hierarchy. Each cluster has a cluster head (similar to the cluster leader in the NICE protocol) and a cluster associate head. The size of a cluster is between k and $3k$, where k is a constant and $k > 3$. ZIGZAG tries to avoid the bottleneck in the NICE protocol by limiting the out-degree of the peers in the distribution tree. The cluster associate head, instead of the cluster head, is responsible for forwarding the media content to the cluster it belongs and obtaining the media content from one peer in the upper layer. As a result, the degree of the distribution tree in ZIGZAG is bounded by $6k - 3$.

Both of the above schemes construct a single tree for the media content delivery which causes two problems: First, some peers, for example, the cluster leader in NICE or the cluster head in ZIGZAG, are more important than other peers. Thus it suffers point of failure problem; Second, the uplink bandwidth of the leaf peers is wasted. Next, we discuss two schemes that employ multiple trees to alleviate these problems.

SplitStream [20] is proposed to overcome the unbalanced forwarding load in conventional tree-based approaches and the traffic stoppage in peer failure or sudden departure. The key idea in SplitStream is to split the media content into n stripes and to multicast each stripe using a separate tree. Peers join different trees to receive different stripes, respectively. The goal of SplitStream is to construct a forest of trees such that each peer is an interior node in one tree and a leaf peer in all the remaining trees while minimizing the delay and link stress across the system.

As Fig. 5.13 shows, two trees are constructed to span the same set of peers. Each peer, except the root, receives two stripes and forwards a strip twice. In this way,

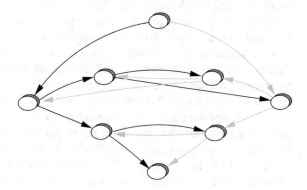

Fig. 5.13 Illustration of multiple distribution trees in SplitStream

the forwarding load is distributed among the peers evenly and a peer failure will cause the data loss of at most one stripe at the downstream peers. When coupled with some media encoding scheme, peers can always reconstruct the media content with the quality proportional to the number of the stripes it receives.

More recently, a scheme to minimize the end-to-end delay in the overlay topology of a peer-to-peer streaming system is proposed in [25]. It assumes that peers have heterogeneous uplink bandwidths and uniform downloading rates. It first formalizes the problem into Minimum Delay Mesh problem (We will use MDM to represent the scheme in the rest of the chapter). Then it proposes a power-based distributed algorithm. *Power* is defined between a child and its parent which is the ratio between the parent's residual uplink bandwidth and the delay from the child to the root through the parent. The parent with higher power is preferred since it offers higher bandwidth or less delay. When a new peer joins the system, it sorts the existing peers in a descending order of their powers and chooses the peers as its parents from the beginning until its downloading rate requirement has been satisfied.

The approach proposed in this section is also based on the idea of adopting multiple trees for constructing overlay topology for peer-to-peer streaming system. However, there are two critical differences distinguishing it from the existing work. First, it does not require the tree to span all the peers as required in SplitStream. Second, it does not require the downloading rates of the peers to be uniform as required in MDM. Thus, it is much more general than the previous schemes. The basic idea is to model the peer-to-peer streaming system by a graph G on which maximum edge disjoint trees can be found. As a result, our approach can be used for peer-to-peer streaming systems where the downloading rate requirement is heterogeneous and the uplink bandwidth of the peers is limited.

5.3.1.1 Problem Formalization

In a typical peer-to-peer streaming system, a streaming server hosts the media content and all the peers requesting the media content retrieve the streaming data from the server directly or from other peers. As peers can be any computers connected to the network, different peers have different access link bandwidths, which is called link heterogeneity. Link heterogeneity is a common phenomenon in today's Internet. Any practical peer-to-peer streaming system should take link heterogeneity into consideration. In this section, we consider the system where each peer has asymmetric access links, which is common for the ISP providers nowadays. Each peer downloads the media content through *downlink* at a speed upper bounded by the downlink bandwidth (BW_{down}), and uploads the media content through *uplink* at a speed upper bounded by the uplink bandwidth (BW_{up}). The playback quality of a peer is determined by the downloading rate of the peer (here we assume that the peer always prefers to download the media content at a higher bit rate if the downlink bandwidth permits. In the case that the downlink bandwidth is higher than the required bit rate, we change the downlink bandwidth to the required bit rate. We adopt the network model where all the peers are connected to a high speed

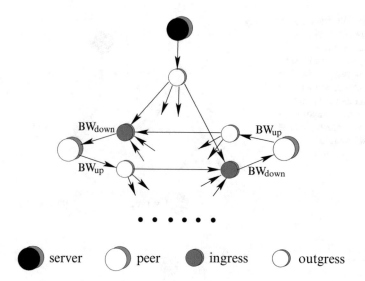

Fig. 5.14 Illustration of the network model for heterogeneous peer-to-peer streaming systems

network core with their access links. The bandwidth bottleneck only lies on the network edge, while the network core has sufficient bandwidth to transmit data among different peers simultaneously. We formulate the network model into a graph $G = (V, E)$ as shown in Fig. 5.14. For each peer, there are two artificial nodes which are used to model the downlink and uplink, respectively. The link between the ingress artificial node is the downlink with bandwidth BW_{down}. The link between the outgress artificial node is the uplink with bandwidth BW_{up}. The source node representing the server has only an outgress node and the bandwidth of the link between the source node and its outgress node equal to the uplink bandwidth of the server. Each outgress node is connected to all the ingress nodes with sufficiently large bandwidth except the ingress node corresponding to the same peer as the outgress node.

As different peers have different downlink bandwidths, it is preferred that each peer is able to download the media content at its maximum downloading rate, i.e., BW_{down}. In peer-to-peer streaming systems, peers download the media content not only from the server, but also from other peers. A mesh overlay is constructed to represent the data flow among peers. To achieve the best playback quality and utilization of access link bandwidth, the overlay topology must be constructed carefully to meet the streaming quality requirement and avoid bandwidth wasting.

Here are some issues that need to be considered in the overlay topology construction. First, the uplink bandwidth of the server is the upper bound of the downloading rate of all the peers. This is clear because a peer cannot download the media content at a speed higher than the server can provide. Therefore, if the downlink bandwidth of a peer is greater than the uplink bandwidth of the server, its downlink bandwidth cannot be fully utilized. In the network model graph G, we

Fig. 5.15 Illustration of the
necessary but insufficient
condition for overlay
topology for peer-to-peer
streaming systems. (a) The
bandwidth configuration of
the server and the peers.
(b) The overlay topology to
achieve the maximum total
downloading rate

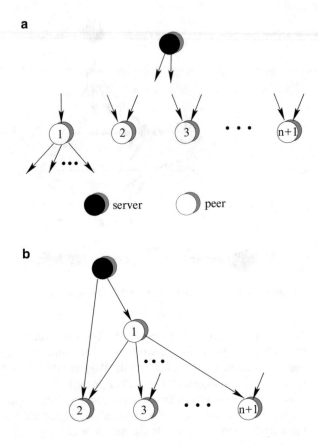

change the downlink bandwidth which is greater than the upper bound to the upper
bound. This change will not alter the resulting overlay topology. Another necessary
condition for all the downlink bandwidth to be fully utilized is that the total uplink
bandwidth is greater than the total downlink bandwidth. This is a necessary but not
a sufficient condition. Its insufficiency can be explained by the following example
as shown in Fig. 5.15.

Suppose there are a total of $n + 1$ peers in the system as shown in Fig. 5.15a.
The uplink bandwidth of the server is 2. Peer 1 has a downlink bandwidth of 1 and
an uplink bandwidth of $2n - 1$. The remaining n peers have downlink bandwidth
of 2 and 0 uplink bandwidth. Therefore, the total uplink bandwidth is $2n + 1$, and
the total downlink bandwidth is also $2n + 1$. Note that although the total downlink
bandwidth of the n peers is the same as the total uplink bandwidth of the server and
the peers, it is impossible to construct an overlay topology which can transmit the
media content to all the peers at their maximum downloading rates. In the best case
as shown in Fig. 5.15b, only two peers (peer 1 and peer 2) can download the media
content at their maximum downloading rates, and all the remaining peers can only
download the media content at half of its maximum downloading rate.

It can be seen that the downlink bandwidth cannot be fully utilized even if there is enough uplink bandwidth. The goal is to maximize the utilization of downlink bandwidth given the network model graph G. The overlay topology plays a critical role in the utilization of downlink bandwidth. In the previous example, if the server dedicates all its uplink bandwidth to a peer with a downlink bandwidth of 2, then all the peers except that peer cannot receive any media content through the server or other peers. Given a network model graph G and an overlay topology, the downloading rate of a peer is equal to the minimum cut between the source node and the peer. Here the overlay topology includes not only the links between peers, but also the bandwidth allocation on the links.

According to the above discussion, the problem of maximizing the utilization of downlink bandwidth can be formalized as follows: Given a peer-to-peer streaming system G, construct an overlay topology such that the sum of the minimum cuts between the source and each peer is maximized.

The problem formalization does not provide a clue in solving the problem. We notice that although the overlay topology is a mesh, the media content delivery is actually through multiple trees. If we divide the stream into multiple unit substreams such that one unit substream occupies one unit of bandwidth, the overlay topology can be decomposed into multiple trees each of which represents a substream. Then the downloading rate of a peer is equal to the number of trees connected to the peer. If we define the length of a tree as the number of its nodes minus one, the total downloading rate of a substream is equal to the length of the tree. Therefore, the problem can be transferred to:

Given G, find a set of edge disjoint trees rooted at the source node such that the sum of the length of the trees is maximum.

We call it maximum downloading rate (MDR) problem. To show the hardness of the problem, we first take a look at the Steiner tree packing problem.

Steiner tree packing problem: Given a graph G, find the maximum number of edge disjoint subgraphs that connect a given set of nodes.

It has been shown that Steiner tree packing problem is an NP hard problem [28]. The difference between the MDR problem and the Steiner tree packing problem is that Steiner tree requires each tree spans all the nodes in the given set, while in MDR problem, the trees are allowed to span part of the peers as long as the total tree length is maximum. In other words, the Steiner tree packing problem is a special case of the MDR problem. Therefore, the hardness of the MDR problem is no less than that of the Steiner packing tree problem, thus MDR is NP hard.

5.3.1.2 The Greedy Heuristic Algorithm

Due to the NP hardness of the MDR problem, as a starting point, in this section we propose a greedy heuristic algorithm to find the maximum number of edge disjoint trees given the network topology G. It is a centralized algorithm which requires the complete information on the peers and their access link bandwidth. As we know, in practice, this information is dynamic and usually cannot be obtained in advance.

We will further propose a distributed algorithm to handle the dynamics of peers in the next section. Nonetheless, the centralized heuristic algorithm serves as the foundation for the distributed algorithm and the benchmark when we evaluate the performance of the distributed algorithm in Sect. 5.3.1.7.

The basic idea of the greedy heuristic algorithm is to pick out a maximum edge disjoint tree from graph G one by one until there is no such a tree. Before we apply the algorithm to graph G, we need to modify the graph by replacing each link of bandwidth b with b parallel links each of which has bandwidth 1. This modification will not change the result of the algorithm. Therefore, each tree represents a substream with bit rate of 1. The receiving bit rate of a peer is equal to the number of trees the peer is on. Each tree must be rooted at the source node. To ensure the server transmits as many substreams as possible, the number of children of the root is limited to one for each tree. Two factors are taken into consideration for the tree construction. First, the height of the tree should be minimized. The height of the tree determines the delay between the source and the peer. To minimize the height of the tree, we should push the peers with higher uplink bandwidths to the source node as close as possible. Second, fairness bandwidth allocation between different trees should be maximized. When a peer receives the media content from several different trees, the delays to the source along the trees may be different. The final end-to-end delay from the source to the peer is determined by the worst case, i.e., the longest delay among all the delays. One way to minimize the difference among these delays is to minimize the difference of the tree heights, as the height of the tree represents the longest end-to-end delay approximately. Fairness here means that the uplink bandwidth should be shared among different trees in a fair manner. As a result, the height difference among different trees is minimized. The fairness is realized by a property of the peer called *fanout*. Fanout is a value which is used to evaluate the forwarding ability of a peer. The fanout of a peer is equal to (uplink bandwidth)/(downlink bandwidth). The larger the fanout, the more children the peer can have. We use fanout as the upper bound on the number of children a peer can have in one tree. The reason is two folds. On one hand, if the number of children is smaller than the fanout, some uplink bandwidth will be wasted definitely even if the downlink bandwidth is filled up with the substreams. On the other hand, if the number of children is more than the fanout for one tree, it means that the number of children is less than the fanout for another tree. This imbalance of uplink bandwidth allocation contradicts the fairness principle.

Table 5.1 lists the pseudo-code of the greedy heuristic algorithm, where the *maximum_tree* sub-procedure is used to find a maximum tree in the residual graph G'.

5.3.1.3   The Distributed Algorithm

The greedy heuristic algorithm works well if the information of the peers and their access link bandwidths are given in advance. However, in practice, peers join or leave the peer-to-peer system frequently which is known as *churn*. Due to

Table 5.1 Greedy heuristic
algorithm

> **Greedy heuristic algorithm**
> **Input**: graph G.
> **Output**: a set of maximum trees.
> **Begin**:
>   **Foreach** peer v in G
> $fanout(v) =$
>   uplink bandwidth/downlink bandwidth;
>   Put peers in list l in a descending order of fanout;
> $G' = G$;
> $T = $ maximum_tree(G', l);
>   **While** T != NULL
>   output = output+T;
> $G' = G'$-T;
> $l = l$-peers with no residual downlink bandwidth;
> $T = $ maximum_tree(G', l);
> **End**
>
> **Maximum_tree(G, l)**
> **Input**: graph G, list l.
> **Output**: a maximum tree T.
> **Begin**:
> $v = $ the first peer in l;
> $pointer = $ the second peer in l;
>   **While** $pointer$ has not reached the end of list l
>   **Do**
>   **If** $pointer$ has residual downlink bandwidth
>   Connect v with peer $pointer$;
> $pointer = pointer + 1$;
>   **Until** peer v is connected to $fanout(v)$ peers
>   or $pointer$ reaches the end of list l;
> $v = v + 1$;
> **End**

the frequent churn rate, it is impossible for the server to run the greedy heuristic algorithm every time when there is a peer joining or leaving. When a new peer joins the system, it should be grafted to the system in an efficient and distributed fashion. The topology mismatch is another issue when we design a practical peer-to-peer streaming system. Here mismatch means that two nodes close to each other in overlay topology are far away in physical network topology or vice versa. The topology mismatch should be minimized to reduce network bandwidth consumption and end-to-end delay. As peers join the system in a random order, the mismatch between the overlay topology and the physical network topology may become larger and larger. A practical peer-to-peer streaming system should be able to adapt itself to the changing overlay topology by alleviating the mismatch. In this section, we propose a distributed algorithm which can handle the frequent peer join/leave in a distributed fashion and adjust the topology to alleviate the mismatch.

5.3.1.4   Peer Joining

We assume that the server is well-known whose IP address is known to all the peers by some address translation service such as DNS. The server maintains a partial list of existing peers in the system and their IP addresses.

The order of the peers in the list is determined by two factors: fanout and end-to-end delay. As we discussed earlier, fanout represents a peer's forwarding ability. A peer with a larger fanout can have more children, so we need to move the peers with larger fanout values to the server as close as possible in the tree. Similarly, the end-to-end delay between a peer and the server represents the distance between the peer and the server. To reduce the mismatch, it is reasonable to move the peers with shorter end-to-end delay closer to the server. We use a tunable parameter α to control the weights of these two factors. We let $level = \alpha * fanout + (1 - \alpha)/end-to-enddelay$, where $level$ is a variable suggesting the position of the peer in the trees approximately. The larger the value of $level$, the closer the peer should be put to the server in the tree. The partial list maintained in the server is in an ascending order of $level$ to facilitate the join procedure. We will evaluate the impact of the tunable parameter on the system performance in Sect. 5.3.1.7.

When a peer wants to receive the media content, it initiates a join process by measuring the delay between the server and itself. Then it sends a JOIN request with the calculated value of $level$ to the server. The server will respond with a list of peers which are picked in the partial list. The picked peers have similar $level$ values as the joining peer. Compared to the schemes in which the server responds with a random list of peers, the topology constructed will converge to the optimal topology much faster, as the new peer is already put at a near-optimum position during the join procedure and less topology adjustment is needed afterwards. The new peer will establish connections with the peers in the list in a descending order of their values of $level$ until its downlink bandwidth is filled up.

It should be pointed out although the server is responsible for bootstrapping the peers, it will not be the bottleneck of the system, because once each peer receives the list of peers, it communicates directly with the peers for overlay topology construction and media content dissemination.

5.3.1.5   Peer Leaving

We use simple peer leaving process to accommodate resilience and alleviate the impact of churn. When a peer leaves the system, we do not reconfigure the topology which will cause a lot control overhead and sometimes even streaming stoppage. Instead, we let the disconnected peers to simply rejoin the system.

There are two types of peer leaving: friendly or abruptly. Friendly leaving means that the leaving peer initiates a leaving process so that the system is aware of its leaving and can make necessary updates accordingly. Abruptly leaving means that the leaving peer leaves the system without any notification, mainly due to link crash or computer crash.

For the friendly leaving, the leaving peer will initiate a leaving process by sending LEAVE messages to both of the server and its neighbors (parents and children). The leaving of the peer will cause the peers which are its descendants in the trees to disconnect from the trees and lose some substreams. To reconnect these peers to the trees, the children of the leaving peer will initiate the join procedure like a new peer. For the parents of the leaving peer, they will free the uplink bandwidth used by the leaving peer for the future use. Here the server acts as a connection "hub" for the peers that are connected to the leaving peer. This may increase the processing burden on the server temporarily. Nevertheless, it can achieve strong robustness with little control overhead. For example, it can handle concurrent peer leavings. When the server receives the LEAVE message, it will check whether the peer is on the partial list and remove it from the list if it is on the list.

For the abruptly leaving, peers send HELLO messages to its neighbors periodically and maintain a HELLO timer for each neighbor. Receiving a HELLO message triggers a reset of the corresponding HELLO timer. The neighbors detect the abruptly leaving by the timeout of the HELLO timer. After detecting the leaving of the peer, the parents and the children will perform similar operations to that in the friendly leaving. Besides, the parents will send LEAVE messages to the server on behalf of the leaving peer so that the server can update the list.

5.3.1.6 Topology Adjustment

Peers join and leave the system in a random manner. According to the join procedure, a new peer is always a descendant of the existing peers after it joins the system. As a consequence, the overlay topology is changing as the peers join and leave and is dependent on the order of the peers joining and leaving. Due to the randomness of the order of peer joining and leaving, the overlay topology may be far from the one constructed by the greedy heuristic algorithm. We propose a topology adjustment procedure to help the peer-to-peer system handle the dynamic peer joining and leaving.

The intuition behind the adjustment procedure is to locate the peers whose positions do not match their *level* values and move these peers to a proper position. It is composed of three steps. (1) Step 1. The peer (We use v to denote the peer in the rest of this section) sends a REQUEST message upstream along the tree towards the root. The REQUEST message includes the value of *level* of v and a hop counter. Each peer receiving the REQUEST message will forward it to its parent and add 1 to the hop counter. The REQUEST message will stop when it reaches the root. (2) Step 2. Each peer receiving the REQUEST will compare the value of *level* in the message with its own value of *level*. If its own *level* is less than that in the message, it will send a GRANT message back to v. The GRANT message contains its own value of *level*, its parent, the residual uplink bandwidth of its parent and the value of the hop counter. (3) Step 3. If v does not receive any GRANT message, it does nothing. If v receives one or more GRANT messages, it will choose one peer to be its new parent based on the following rules: sort the peers in a descending order of their hop counts; select the first peer whose parent has non-zero residual uplink bandwidth and take its parent as the new parent of v. The peer v and the subtree rooted at v will become a subtree of the new parent. If there is no such a peer whose parent has non-zero residual uplink bandwidth, the peer checks itself to see if there is any residual uplink bandwidth. If no, the peer does nothing. If yes, the peer v will select the peer with the largest hop count and take its parent as the new parent. As there is no residual uplink bandwidth of the parent, one child of the parent will be replaced by v. The old child will become a child of v. The rationale behind the rules is to move the peer as close to the root as possible while minimizing the disturbance to the existing tree structure.

5.3.1.7   Performance Evaluations

We have implemented the proposed scheme in NS-2 [34]. The network topologies used in the simulations are random transit-stub network topologies generated by GT-ITM software [11]. Peers are selected randomly from the stub networks and the bandwidth of the links in the transit network is set to be sufficiently high (1000 Mbps in the simulations).

We compare the scheme with the MDM algorithm proposed in [25], as it is the closest work to ours in the sense that both of the algorithms try to optimize the overlay topology for peer-to-peer media streaming systems. The main difference is that the scheme considers heterogeneous downloading rates while MDM assumes a uniform downloading rate.

The simulation adopts following three performance metrics:

Satisfaction. Even though the overlay topology is carefully constructed, it is inevitable that a peer may not receive the media content at its preferred bit rate. We use *satisfaction* to evaluate the extent the peer is satisfied with its received media streaming bit rate. A peer's satisfaction is defined as the ratio of the received media streaming bit rate to its downlink bandwidth. In the simulation, we use the average satisfaction of all peers to represent the satisfaction of the system.

End-to-end delay. This performance metric is used to evaluate the end-to-end delay between the source and the peers. The end-to-end delay is measured along the overlay links towards the source instead of the physical links. Since there are usually multiple overlay paths towards the source, we use the delay of the longest path as the end-to-end delay of the peer. Again, we average the end-to-end delay of all peers in the simulations. Since the time is the virtual simulation time in NS2, the time unit is virtual as well.

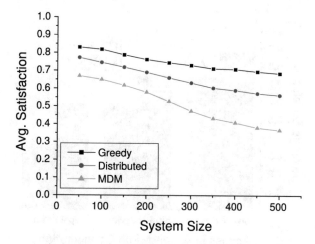

Fig. 5.16 Average satisfaction under different system sizes

Link stress. Link stress is defined as the number of copies of the same message transmitted through the same link. It is a performance metric that only applies to an overlay network due to the mismatch between the overlay network and the physical network. We use it to evaluate the effectiveness of the topology adjustment and the efficiency of the system.

Satisfaction. We first compare the average satisfaction of peers under different sizes of the peer-to-peer streaming system. We use the number of peers in the system to represent the size of the system.

Figure 5.16 shows the average satisfaction as we change the system size. We can observe that the greedy heuristic algorithm demonstrates the best satisfaction and the distributed algorithm outperforms MDM by about 30 % on the average. It suggests that proposed scheme can make more efficient use of the uplink bandwidth of peers than MDM. With the increase of the system size, the average satisfaction drops for all the three schemes. This is due to the fact that when the trees become large, the wasted uplink bandwidth of the leaf peers increases as well even though multiple trees are employed. From the trend of the curves, we can see that the scheme is more stable then MDM with the increase of system size. The decrease of the satisfaction of the distributed algorithm is about 26 % while that of MDM is about 45 %.

To examine the system performance under different access link bandwidth constraints, we let the ratio between the total uplink bandwidth and the total downlink bandwidth equal to 4, 2, and 1, respectively. Figure 5.17a shows the simulation results. We can see that the average satisfaction of the distributed algorithm is always higher than that of MDM. Moreover, the average satisfaction of the distributed algorithm when the total uplink bandwidth equals to the total downlink bandwidth is higher than that of MDM when the total uplink bandwidth is two times of the total downlink bandwidth. It indicates that the proposed scheme performs well especially when the total uplink bandwidth is limited.

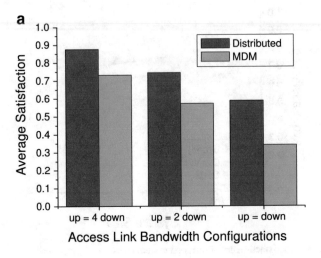

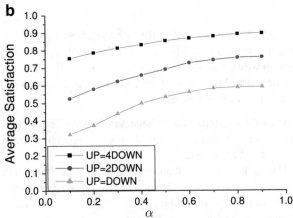

Fig. 5.17 Average satisfaction evaluation. (**a**) Average satisfaction under different access link bandwidth configurations; (**b**) Average satisfaction under different values of α

The tunable parameter α is another factor that affects the system performance. Figure 5.17b shows the average satisfaction under different values of α and different access link bandwidth configurations. We can see that the average satisfaction increases with the increase of α. When α is small, end-to-end delay enjoys more weight in the overlay topology construction. The resulting topology tends to have a shorter end-to-end delay at the expense of the satisfaction. The slope, i.e., the increasing rate of the average satisfaction, is greater when α is small than that when α is large. This can be used as a system design guide to find an optimal α value for both satisfaction and end-to-end delay.

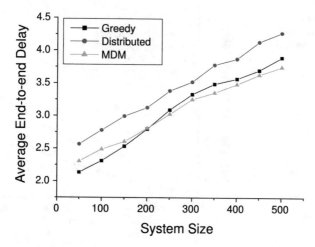

Fig. 5.18 Average end-to-end delay under different system sizes

End-to-End Delay. The end-to-end delay is an important performance metric for peer-to-peer media streaming systems. As we can see later, sometimes end-to-end delay and satisfaction are two conflicting performance metrics and the best solution is a tradeoff between these two metrics.

Figure 5.18 shows the end-to-end delay under different system sizes. We can see that with the increase of the system size, the end-to-end delay increases as well due to the large tree heights. When the system size is small, the greedy heuristic algorithm achieves the shortest end-to-end delay. As the system size becomes large, the end-to-end delay of MDM is shorter than that of other two schemes. The reason is that MDM is a scheme focused on minimizing the end-to-end delay. The advantage of MDM is more obvious when the system size is large.

Now we investigate the impact of the parameter α on the end-to-end delay. As shown in Fig. 5.19, the end-to-end delay increases with the increase of α. The rationale behind this is similar to that of the results for satisfaction. As we put more weight on *fanout* when we construct the overlay topology, the resulting end-to-end delay becomes longer. When the uplink bandwidth constraint is tight, the increase of the end-to-end delay is faster compared to that when the uplink bandwidth constraint is loose. This attributes to the larger tree height due to the limited uplink bandwidth when the uplink bandwidth constraint is tight.

Link Stress. Link stress is an indicator of the efficiency of the overlay topology. Higher link stress will cause higher delay and bandwidth wasting.

We first look at the link stress under different system sizes as Fig. 5.20 shows. With the increase of the system size, the link stress increases as well. The greedy heuristic algorithm achieves the least link stress thanks to the complete peer information before the overlay topology construction. The link stress of MDM is slightly higher than that of the distributed algorithm. This is because that although

Fig. 5.19 Average
end-to-end delay under
different values of α

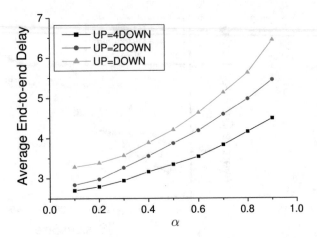

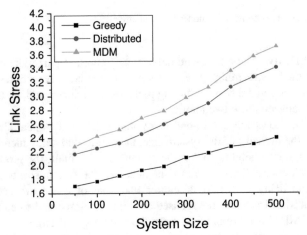

Fig. 5.20 Link stress under different system sizes

the proposed distributed algorithm considers heterogeneous download rates while
MDM only considers a uniform downloading rate, many trees constructed in the
distributed algorithm have smaller heights as they do not span all the peers. As a
result, the probability that a packet passes the same physical link is reduced and the
link stress is reduced as well.

Figure 5.21 shows the link stress when we tune the parameter α. The impact of
parameter α on link stress is notable. With the increase of α, link stress increases
remarkably. As mentioned earlier, a larger α means shorter end-to-end delay. While
the end-to-end delay is a good approximation of physical distance between two
peers. When the end-to-end delay is reduced, the mismatch between the overlay
network and physical network is reduced as well. The impact of the uplink
bandwidth constraint on link stress is small. When the uplink bandwidth constraint

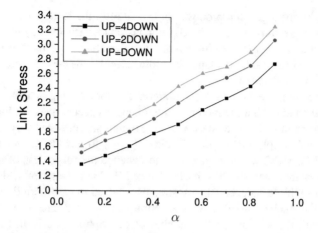

Fig. 5.21 Link stress under different values of α

is loose, the total download bandwidth is increased. However, this increase will not cause the increase of the link stress, because it becomes more likely to find a peer nearby to exchange the media content with.

5.3.2   Adaptive Network Coding for Peer-to-Peer Media Streaming Systems

The paper in [7] gives a theoretical analysis on how network coding can improve the performance of peer-to-peer streaming systems. A random network coding scheme was proposed for peer-to-peer media streaming systems in [36]. The scheme focuses on utilizing random network coding to simplify the scheduling algorithm for peers to exchange the media content. However, it does not consider peers' heterogeneity in peer-to-peer streaming systems. A network coding scheme based on layered encoding was considered in [33, 49] for media streaming systems. The scheme proposed in [33] was designed for media streaming multicast networks while the scheme proposed in [49] was designed for peer-to-peer streaming systems. Layered Coding (LC) [35] is a multimedia encoding technique which encodes the media content into a base layer and multiple ordered enhancement layers. The receiver can reconstruct the media content if it receives the base layer and a subset of the enhancement layers. It is a cumulative encoding technique in the sense that an enhancement layer can be used for decoding only if the base layer and all the enhancement layers before it are received by the receiver. The quality of reconstructed media content is proportional to the layers used. In [33], the receivers are divided into several groups based on their max-flow values to the source. Then a subgraph is constructed for each group of receivers. Given a subgraph, a deterministic linear network coding scheme is determined by the algorithm proposed in [5]. In [49], the authors used the overlay construction algorithm proposed in [31] to construct a basic overlay network. Then layered meshes are constructed for layered media content. Each mesh is responsible for one layer of the media content. Peer join different meshes to receive different layers of the media content.

In this section, we introduce an adaptive network coding scheme which can optimize the bandwidth utilization in a heterogeneous peer-to-peer media streaming system. The main differences between the existing schemes and this scheme are: First, it uses Multiple Description Coding (MDC) to encode the media content instead of LC. MDC was first proposed to enhance the robustness of multimedia data over unstable channels. The basic idea of MDC is to fragment a single media stream into multiple independent substreams referred to as descriptions. In order to decode the media stream, any description can be used. However, the quality improves with the increase of the number of descriptions received. Here we use MDC to encode the media content into multiple stripes which are equally important when used to reconstruct the media content. The quality of the reconstructed media content is proportional to the number of stripes used. This leads to a great advantage over LC in the sense that the peer can always reconstruct the media content as long as it receives one or more stripes. While in LC, if the base layer is missing, the media streaming has to stop. If an enhancement layer is missing, all the enhancement layers after it are of no use which causes bandwidth wasting. Second, since most peers are individual computers which are connected to the Internet through access links, peers may have different bandwidth in practice. The scheme aims at such practical networks by considering asymmetric access links where uplink bandwidth and downlink bandwidth are bounded by a link capacity. This is common for the ISP providers nowadays. Third, we use the overlay construction algorithm described in the previous section to construct the overlay network. It is optimized for heterogeneous peer-to-peer media streaming systems. While other schemes usually adopt an existing general purpose overlay topology construction algorithm. For example, the scheme in [49] used the algorithm proposed in [31].

The scheme first encodes the media content into multiple stripes using MDC technology [15]. Then peers subscribe to a subset of these stripes based on their downlink bandwidths. Network coding is performed within one stripe. The playback quality of a peer is proportional to the number of stripes it subscribes to. The overlay topology construction is also tailored to optimize the bandwidth utilization of the peers' access links. Although the scheme is designed for live peer-to-peer media streaming systems, the result can also be applied to peer-to-peer content delivery or file downloading systems.

5.3.2.1 Problem Formalization

Without loss of generality, the topology of a peer-to-peer streaming system can be modeled as a multicast network which can be represented by a directed graph $G = (V, E, C)$ where V is the set of network nodes and E is the set of links each of which connects two nodes. Each link can be represented by an ordered node pair (v_1, v_2) where $v_1, v_2 \in V$. v_2 is called the head of the link and v_1 is called the tail of the link. The messages can only be transmitted from v_1 to v_2. C is a real non-negative function $C : E \to R^+$ which maps each link e to a real non-negative number $C(e)$ which is the transmission capacity of the link. The media content is generated at a source node $s, s \in V$ and flows to a set of receivers $R, R \subseteq V$.

With the help of MDC, the media content is encoded into multiple stripes at the source node before being sent out. In our model, we encode the media content into k stripes each of which has the same bit rate b. The receiver can reconstruct the media content with any subset of the k stripes. The playback quality of a receiver is proportional to the number of stripes it receives. Without network coding, MDC is realized by finding multiple disjoint multicast trees spanning the source and the receivers. Each tree is responsible for one stripe. With network coding, the bandwidth can be further utilized by applying network coding to the flows in the same stripe. It is possible to encode flows from different stripes. However, it requires decoding in the relay nodes in addition to the receivers. Therefore, due to the complexity, in this section we do not consider network coding between different stripes.

Suppose each stripe is distributed to the receivers through a subgraph (Given a graph $G = (V, E, C)$, a subgraph G' can be defined as $G' = (V', E', C')$ where $V' \subset V, E' \subset E, C'(e') <= C(e'))$. To apply network coding to the flows within one stripe, the subgraph must be a mesh instead of a tree. This implies that we should divide the stripe to multiple flows and transmit these flows along different paths to a receiver in order to increase the probability for nodes to perform network coding. The receivers subscribe to one stripe if they want to receive the correspondent media content. The number of stripes a receiver subscribes to is upper bounded by the bandwidth of its downlink divided by b. By subscribing to different numbers of stripes, the utilization of heterogeneous downlink bandwidths can be maximized.

Our goal is to maximize the total receiving rate of all the receivers. Now the problem is transformed to how the receivers subscribe to the stripes such that the total receiving rate is maximized. Assuming that the subset of stripes receiver r subscribes to is $F(r)$, the problem can be formalized as a mathematical optimization problem as follows:

$$\text{maximize} \sum_{i=1}^{|R|} |F(r_i)| \tag{5.12}$$

subjectto

$$\sum_{\text{head}(e)=v} x_i^j(e) - \sum_{\text{tail}(e)=v} x_i^j(e) = \sigma_i^j(v),$$

$$\forall\, v \in V, r_i \in R, j \in F(r), \tag{5.13}$$

$$\sum_j \phi^j(e) <= c(e), \tag{5.14}$$

where

$$\sigma_i^j(e) = \begin{cases} -b \text{ if } v = s \\ b \text{ if } v = r_i \\ 0 \text{ otherwise} \end{cases} \qquad (5.15)$$

$$\phi^j(e) = \text{Max}_i\{x_i^j(e)\}. \qquad (5.16)$$

In the above, $x_i^j(e)$ is the flow rate on link e for receiver r_i on stripe j. As each stripe has the same bit rate, to maximize the total receiving rate is equal to maximize the number of total stripes the receivers subscribe to. Equation (5.2) means that only the source node can generate flows and only the receivers can consume flows, while all the remaining nodes perform relaying. Equation (5.3) means that flows from different stripes can not share the bandwidth, and the summation of their bit rates can not exceed the link capacity. Equation (5.4) means that all the stripes have a constant bit rate b. Finally, Equation (5.5) means that the flows from the same stripe can share the bandwidth.

In addition to achieving the maximum throughput, we also want to maintain fairness among receivers. Fairness is defined as follows: the receiver with a larger max-flow value from the source node will receive no less stripes than the receiver with a smaller max-flow value. Then the optimization problem can be rewritten by adding one more constraint:

$$|F(r_i)| \le |F(r_j)| \text{ if } \text{max-flow}(r_i) \le \text{max-flow}(r_j).$$

5.3.2.2 Adaptive Network Coding for Heterogeneous Peer-to-Peer Media Streaming Systems

The solution to the above mathematical optimization problem requires centralized processing with all the topology and bandwidth information available. In a large scale distributed system such as the Internet, it is not scalable to deploy such a centralized algorithm. To accommodate scalability, it is necessary to develop a distributed algorithm implemented by network protocols. However, the problem formalization provides some insights and guidelines leading to the distributed solution.

In this section, we propose a distributed adaptive network coding construction scheme based on the discussion in the previous section.

Overlay Topology Construction. The first step is to construct an overlay network spanning the peers over which our adaptive network coding scheme can be applied. This can certainly be achieved by using some existing algorithms in the literature, such as those in [19, 31]. However, as most of existing algorithms are of general purposes, we will use the previously introduced overlay topology construction algorithm tailored for heterogeneous peer-to-peer media streaming systems for efficiency purpose.

We still use fanout to evaluate the forwarding ability of a peer. The topology construction heuristic is based on the observation that the probability that a peer with a higher fanout relays media content to a peer with lower fanout is higher than the reverse. Fanout can be roughly considered as the number of children in media content relay. Therefore, if a peer has a higher fanout, it tends to have more children to relay media content, which increases the probability that it relays media content to other peers. The reason we use fanout instead of uplink bandwidth as the metric of the peer's forwarding ability is to minimize the average end-to-end delay from the source to the receivers. If we track each single message (either in its original form or in an encoded form), it is distributed through a tree. Assigning higher priority to nodes with larger fanout can reduce the average tree height, therefore reduce the end-to-end delay.

We adapted the proposed overlay topology construction algorithm slightly by adding two more constraints.

1. Constraint 1: if the number of outgoing links is equal to $2 * fanout$, no more outgoing links are added;
2. Constraint 2: if the summation of the download bandwidth of a peer's parents is no less than twice of its download bandwidth, no more incoming links are added.

The selection of the ratio value 2 is based on our simulations, which achieves a good tradeoff between system performance and computing complexity.

The pseudo-code of the overlay topology construction algorithm is listed in Table 5.2.

Table 5.2 Overlay topology construction algorithm

```
INPUT: BW_up(i), BW_down(i)
//BW_up(i) and BW_down(i) are upload bandwidth and
//download bandwidth of node i
OUTPUT: overlay topology
BEGIN
  foreach node i
    fanout(i) = BW_up(i)/BW_down(i);
  Sort nodes in a descending order of fanout into list q;
  E = NULL;
  foreach node i in q
    tail = i;
    head = i + 1;
    while head! = NULL
      if ∑_{j,(j,head)∈E} BW_down(j) < 2 * BW_down(head)
        E = E + (tail, head);
      head + +;
      if |E_out(tail)| == 2 * fanout(tail)
        //E_out(tail) = {l, l ∈ E, tail(l) = tail}
        break;
END
```

Peer Joining. The media content is encoded into multiple stripes at the source node. As there is no coding between different stripes, it is possible to decompose the network topology graph G into multiple disjoint subgraphs each of which is corresponding to one stripe. Therefore, when a peer joins the system, it will select some of the stripes to subscribe to.

When a peer wants to receive the media content, it will send a JOIN request to the source node. Upon receiving a JOIN request, the source node initiates a process to determine the maximum flow between the source node and the joining peer. There are many existing algorithms for this problem such as Ford–Fulkerson algorithm proposed in [8] and push-relabel algorithm proposed in [14]. Here we adopt the push-relabel algorithm because it is more efficient and it is a distributed algorithm.

We first give a brief review of the push-relabel algorithm. Given a graph $G = (V, E, C)$, a source node $s, s \in V$ and a destination node $t, t \in V$, push-relabel algorithm can find the maximum flow between s and t. In the push-relabel algorithm, each node is assigned a *height* value and an *excess* value. Height is used to control the flow direction. A flow can only be pushed from a higher node to a lower node between two neighbors. Although the difference between the flow entering a node and the flow leaving a node is zero when the algorithm terminates (except nodes s and t), during the execution of the algorithm, the flow difference may be positive, i.e., the flow entering a node is more than the flow leaving a node. We use *excess* to denote the amount of flow difference which is a non-negative value. The value of a flow pushed between two neighbors cannot exceed the residual bandwidth of the link connecting the two neighbors. Initially, the source node has a height of $|V|$ and the destination node has a height of 0. The height of a node v ($v \neq s, v \neq t$) is the number of hops along the shortest path from s to v. The excess of the source node is infinity, i.e., initially the source node will push flow to its neighbor nodes as much as possible. The excesses of other nodes are 0. There are two operations in the push-relabel algorithm: (1) *push* operation which is to push a flow from a node with a larger height to one of its neighbors with a smaller height. The value of the flow is the minimum of the excess of the node and the residual bandwidth of the link over which the flow is pushed. When a flow is pushed over a link, an artificial link connecting the same pair of nodes is added to the topology. The direction of the artificial link is opposite to the link and the capacity of the artificial link is equal to the value of the flow; (2) *relabel* operation is used to update the height values of nodes when no legal push operation can be done. If all the nodes except nodes s and t have 0 excess, the algorithm terminates and the excess of t is the maximum flow between s and t. Otherwise, pick a node with positive excess and increase its height such that it can push a flow to a neighbor.

The push-relabel algorithm is perfect for a distributed system as the flow is determined gradually and locally between a pair of nodes. Thus, we adopt the push-relabel algorithm in our system. Each link is associated with an information vector which includes the following information: link capacity, the stripes which have flows on the link and the bandwidths the stripes occupy, respectively. The link information is used by a peer to decide which stripe to subscribe to and it is collected when a push operation is performed. In particularly, in the push operation, in addition to pushing the excessive flow from the higher node to the lower node, the algorithm records the corresponding information which is carried along the flow.

When the algorithm terminates, the joining peer should have following information:

1. The paths from the source to the peer,
2. For each link in a path, the stripes which are transmitted over it and the bandwidths occupied by them.

Based on this information, the joining peer selects the stripes to subscribe to. The heuristic rules used for the selection are as follows:

1. Give higher priority to the stripes which have flows in at least one path.
2. Give higher priority to the stripes which occupy more paths.
3. Subscribe to as many stripes as possible.

The first two rules are quite straightforward, because subscribing to the stripes that already have flows in the paths can maximize the utilization of the bandwidth that is already used by the stripes. It is similar to grafting a new receiver to an existing multicast tree in multicast routing. The last rule means that if there is still residual bandwidth after subscribing the stripes based on the first two rules, subscribe to new stripes to fill up the residual bandwidth.

In LC, there are different priorities assigned to different layers as well. For example, the base layer is given the highest priority. However, the priority used here is different from that used in LC. In LC, the priority is mandatory such that the layer with lower priority is useless unless all the layers with higher priorities are received. The priority is used to differentiate different layers. In our scheme, the priority is not mandatory. Peers can choose stripes with lower priority instead of higher priority to encode and every stripe is used to decode at the receivers. Priority is used to improve the bandwidth utilization.

The pseudo-code for the stripe selection algorithm is listed in Table 5.3.

Peer Leaving. When a peer leaves the system, it needs to leave the stripes it subscribes to. It performs the leave procedure repeatedly for each stripe.

To leave a stripe, a peer first checks its position on the subgraph of the stripe. There are two cases:

Case 1: It has no outgoing links. In this case, the leaving peer sends PRUNE message to its parents (maybe more than one) such that they can delete the links from the subgraphs.
Case 2: It has one or more outgoing links. In this case, the peer will notify its children to rejoin the system. Meanwhile, it will send PRUNE message to its parents.

Sometimes, peers may leave the system due to crash. In this case, it is impossible for the leaving peer to notify its neighbors. To solve this problem, peers send HELLO messages to its neighbors periodically and maintain a HELLO timer for

Table 5.3 Stripe selection algorithm

INPUT: path p_i, link e_{ij}, link capacity c_{ij}, stripes f_{ij}^k
         and bandwidth b_{ij}^k
//link e_{ij} is the jth link on path p_i
OUTPUT: stripes to subscribe to
BEGIN
  foreach stripe k
    if bandwidth_is_enough(k)
      Put stripe k into a list l;
  Sort l in a descending order of the number of paths the
  stripe occupies;
  foreach stripe k in l
    if bandwidth_is_enough(k)
      foreach path p_i
        Allocate $b * cap_i^k / \sum_i cap_i^k$ bandwidth to stripe k;
END

bandwidth_is_enough(k)
BEGIN
  foreach path p_i
    foreach link e_{ij}
      $cap_{ij}^k = c_{ij} - \sum_k b_{ij}^k + b_{ij}^k$;
      $cap_i^k = min_j\{cap_{ij}^k\}$;
  if $\sum_i cap_i^k \geq b$
  return TRUE;
  else
  return FALSE;
END

each neighbor. Receiving a HELLO message triggers a reset of the corresponding HELLO timer. The neighbors detect the abruptly leaving by the timeout of the HELLO timer. After detecting the leaving of the peer, the parents and the children will perform the same operations discussed earlier.

Network Coding. We apply random network coding within each stripe. After encoding the media content into k stripes, the source node sends each stripe by dividing it into groups called *generations*. A generation is a unit for network coding. Only messages within the same generation can be encoded together by network coding. The source node first performs random coding for messages in the same generation. The encoded messages are sent to the outgoing links over which the corresponding stripe has flow.

The relay peers perform random coding when receiving the stripes of the media content. As there is no coding between different stripes, peers only mix the flows from the same stripe and the same generation. When a peer receives flows belonging to the same stripe and generation from its parents, it mixes them up with random coefficients generated from a large Galois field. The mixed flow is sent out only to those outgoing links over which the stripe has flow. By doing this, it is guaranteed that a peer will not receive the flow which is mixed with stripes it does not subscribe to.

The peers decode the messages in the same generation once they receive enough linearly independent messages from their parents. It will acknowledge the source node of its successful receiving of the generation.

5.3.2.3   Performance Evaluations

We have implemented the proposed scheme in NS-2 [34]. The network topologies used in the simulations are random transit-stub network topologies generated by GT-ITM software [11]. Peers are selected randomly from the stub networks. The links between the stub networks and the transit network have 8 Mbps bandwidth on average. The bandwidth of the links in the transit network is set to be sufficiently high (10,000 Mbps in the simulations) to simulate the network model discussed in Sect. 5.3.2.2.

We compare our scheme with a recently proposed scheme in [49], called LION, as it is the closest work to ours. We use ANC (Adaptive Network Coding) to represent our scheme in the figures in the following.

The simulation adopts following three performance metrics:

Satisfaction: A peer may not receive the media content at its maximum rate (downlink bandwidth). We use *satisfaction* to evaluate the extent the peer is satisfied with its receiving rate. A peer's satisfaction is defined as the ratio of the received rate to its downlink bandwidth. In the simulation, we use the average satisfaction of all peers to represent the satisfaction of the system.

Resilience: Resilience is a performance metric used to evaluate the ability of the system to handle peers' dynamic join/leave called churn. We let the peers join and leave the system during the time span. We assume that the uptime of a peer follows Poisson distribution [1]. We evaluate the resilience by examining the throughput under dynamic peer join/leave. Throughput is defined as the service the system provides in one time unit. Here the service is the media content received by the receivers. Due to the heterogeneity of the receivers, the receivers are receiving the media content at different rates. We evaluate the volume of the media content received by every receiver in a given time span. The throughput is equal to the summation of the volumes of received media content by all the receivers divided by the length of the time span.

Control Overhead: Control overhead is a performance metric used to evaluate the efficiency of the scheme. It is measured by the number of control packets during the time span. In particular, control overhead includes the packets generated by the push-relabel algorithm and PRUNE messages, etc.

Satisfaction. We first compare the average satisfaction of peers under different sizes of the peer-to-peer streaming system. We use the number of peers in the

Fig. 5.22 Average
satisfaction evaluation. (**a**)
Average satisfaction under
system sizes when using the
overlay topology construction
algorithm in LION; (**b**)
Average satisfaction under
system sizes when using the
overlay topology construction
algorithm in this section

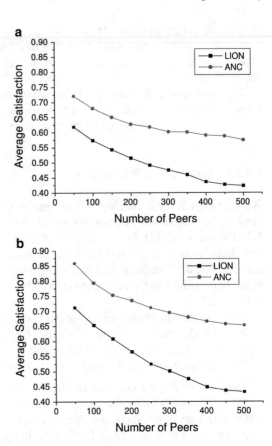

system to represent the size of the system. To investigate the impact of the topology
construction algorithm, we compare the satisfaction using two different topology
construction algorithms: one is proposed in this section and the other is proposed
in [31].

Figure 5.22 shows the average satisfaction as we change the system size. We
can observe that with the increase of the system size, the average satisfaction
drops. This is because that when the number of peers increases, the competition
of the bandwidth resource is more severe. Thus, it is more difficult to allocate
the bandwidth to satisfy every peer's need. For both overlay topology construction
algorithms, our proposed scheme ANC performs better than LION regardless of
the system size. The advantage is greater when the system size is larger, which
suggests that ANC is more scalable than LION. When using the overlay topology
construction algorithm proposed in this section, the satisfaction is increased by
13–20 %.

Resilience. In this subsection, we compare the resilience of the system under
different sizes of the system. Figure 5.23a shows the throughput comparison of
the two schemes with and without churn. We can see that the proposed scheme

Fig. 5.23 Throughput evaluation. (**a**) Throughput under different system sizes when peers join/leave dynamically; (**b**) Throughput under different system sizes and different mean uptimes of peers when peers join/leave dynamically

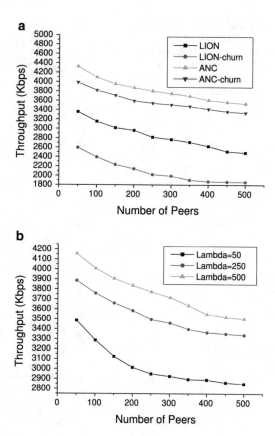

ANC achieves about 18 % higher throughput than LION without churn and about 48 % higher throughput than LION under churn. This can be explained by the fact that when churn occurs, the probability that a peer misses the base layer in LION increases. Without the base layer, it is impossible for a peer to perform decoding. The advantage of MDC over LC leads to the advantage of resilience of ANC over LION.

Since the uptime of the peers follows Poisson distribution, we use λ to denote the mean uptime of peers. Figure 5.23b shows the simulation results when we set the mean uptime to 50, 250, and 500, respectively. The total simulation time is 1000 (in NS-2 time units). We can see that the shorter the mean uptime, the lower the throughput. This is obvious as a shorter mean uptime implies a higher rate at which the peers join or leave the system. When λ is smaller, the throughput is more sensitive to the system size. This is because that a small system size limits the optional paths during churn.

Control Overhead. In this subsection, we compare the control overhead of the system under different sizes of the system. Figure 5.24a shows the curves of the control overhead of the two schemes. With the increase of the system size,

Fig. 5.24 Control overhead
evaluation. (**a**) Control
overhead under different
system sizes; (**b**) Control
overhead under different sizes
and different mean uptimes of
peers

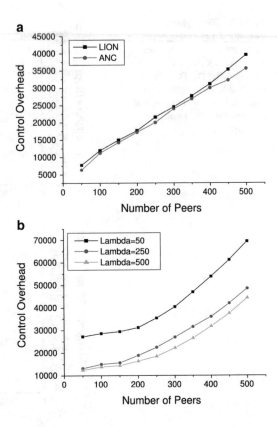

the control overhead increases almost linearly for both schemes. The control
overhead is mainly caused by the join procedure of peers. As the number of
peers increases, the number of control packets increases proportionally as well.
The control overhead of LION is slightly more than that of ANC, because LION
floods packets around the system to find a maximum number of link-disjoint paths
when building the layered mesh. Figure 5.24b shows the impact of λ on the control
overhead. We can see that the control overhead when $\lambda = 50$ is much higher than
that when $\lambda = 250$ or $\lambda = 500$, while the difference between $\lambda = 250$ and $\lambda = 500$
is small.

5.4 Summary

This chapter focuses on network coding in ALM by discussing network coding
applied to peer-to-peer file sharing and peer-to-peer streaming.

PPFEED can serve as a peer-to-peer middleware created within the web services
framework for web-based file sharing applications. Compared to other file sharing
schemes, the advantages can be summarized as follows: (a) Scalability. Files are

distributed through a peer-to-peer network. With the increase of the network size, the total available bandwidth also increases. (b) Efficiency. The linear network coding construction scheme is deterministic and easy to implement. There is no requirement for peers to collaborate to construct the linear network coding assignment on demand. All the peers need is the mapping between the group ID and the encoding function, and this mapping does not change with time. Compared to random network coding, the receiver can always recover the original messages after receiving k different messages and the data dissemination is more efficient as data messages are transmitted through the same overlay link at most once. (c) Reliability. The redundant links can greatly improve the reliability of the system with little overhead. (d) Resilience. Churn is a common problem in overlay networks. By adding redundant links, the negative effect of churn is alleviated. (e) Topology awareness. Simulation results show that the proposed topology clustering scheme can greatly reduce link stress and improve throughput. (f) Heterogeneity support. In case that links have different link capacities, PPFEED can arrange the overlay topology to maximize the utilization of each peer's link capacity.

Peer-to-peer streaming systems are addressed in two steps. First, an optimized overlay topology scheme for live peer-to-peer streaming systems is introduced. One of the merits of the scheme is that it can make efficient use of the uplink bandwidth of the peers and satisfy the heterogeneous downloading rate requirements of the peers as much as possible. Compared to other overlay topology construction schemes, the scheme can handle both heterogeneous uplink bandwidth and hetero-geneous downlink bandwidth at the same time. Peers with different downloading bandwidths can receive the media content at different rates without bandwidth wasting. Besides, the distributed algorithm constructs an adaptive overlay topology which can adapt itself to the changing peers such that the end-to-end delay and link stress are minimized. Simulation results show that the proposed scheme outperforms MDM by about 30 % with respect to the average peer satisfaction. In addition, the proposed scheme achieves less link stress than MDM.

Thereafter, an adaptive network coding scheme for peer-to-peer media streaming systems is introduced. Compared to other peer-to-peer media streaming schemes, the scheme has the following advantages: (a) Heterogeneity support. As most peers are individual computers connected to the Internet through heterogeneous access links, our scheme can maximize the bandwidth utilization of access links and therefore maximize the total throughput of the system. (b) Resilience. Churn is a common problem in peer-to-peer networks. With the help of MDC, peers can reconstruct the media content with a subset of the stripes. (c) Scalability. Media content is distributed through a peer-to-peer network. With the increase of the network size, the total available bandwidth also increases.

References

1. H. Balakrishnan, D. Liben-Nowell, D. Karger, Analysis of the evolution of peer-to-peer systems, in *Principles of Distributed Computing*, July 2002
2. B. Bhattacharjee, S. Banerjee, C. Kommareddy, Scalable application layer multicast, in *Proceedings of ACM SIGCOMM 2002*, Aug 2002
3. Bittorrent, http://www.bittorrent.com, 2004
4. B. Botev, D. Xu, M. Hefeeda, A. Habib, B. Bhargava, Promise: peer-to-peer media streaming using collectcast, in *Proceedings of the 11th ACM international conference on Multimedia*, Nov 2003
5. P. Chou, M. Effros, S. Egner, K. Jain, S. Jaggi, P. Sanders, L. Tolhuizen, Polynomial time algorithms for multicast network code construction. IEEE Trans. Inf. Theory **51**, 1973–1982 (2005)
6. Y. Cui, K. Nahrstedt, Layered peer-to-peer streaming, in *Proceedings of ACM NOSSDAV 2003*, 2003
7. C. Feng, B. Li, On large-scale peer-to-peer streaming systems with network coding, in *Proceedings of the 16th ACM International Conference on Multimedia*, Oct 2008
8. L. Ford, D. Fulkerson, Maximal flow through a network. Can. J. Math. **8**, 399–404 (1956)
9. C. Fragouli, E. Soljanin, *Network Coding Applications* (Now Publishers Inc, 2008) Hanover, MA 02339
10. Gnutella protocol development, the Gnutella v0.6 protocol. Available: http://rfc-gnutella. sourceforge.net/developer/index.html, 2003
11. GT-ITM, http://www.cc.gatech.edu/projects/gtitm/ (College of Computing, George Institute of Technology, 2000)
12. C. Gkantsidis, P.R. Rodriguez, Network coding for large scale content distribution, in *IEEE INFOCOM 2005*, Mar 2005
13. V. Goebel, K. Skevik, T. Plagemann, Evaluation of a comprehensive P2P video-on-demand streaming system. Comput. Netw. **53**(4), 434–455 (2009)
14. A.V. Goldberg, R.E. Tarjan, A new approach to the maximum flow problem, in *Proceedings of the 18th Annual ACM Symposium on Theory of Computing*, 1986, pp. 136–146
15. V.K. Goyal, Multiple description coding: compression meets the network. IEEE Signal Process. Mag. **18**(5), 74–93 (2001)
16. T. Ho, D.S. Lun, *Network Coding: An Introduction* (Cambridge University Press, Cambridge, 2008)
17. K. Hua, D. Tran, T. Do, Zigzag: an efficient peer-to-peer scheme for media streaming, in *Proceedings of IEEE INFOCOM 2003*, Apr 2003
18. R.M. Karp, S. Ratnasamy, M. Handley, S. Shenker, Topologically-aware overlay construction and server selection, in *IEEE INFOCOM 2002*, June 2002
19. A-M. Kermarrec, M. Castro, P. Druschel, A. Rowstron, Scribe: a large-scale and decentralised application-level multicast infrastructure. IEEE J. Sel. Areas Commun. (Special Issue on Netw. Support Multicast) **20**(8), 1489–1499 (2002)
20. A.-M. Kermarrec, A. Nandi, A. Rowstron M. Castro, P. Druschel, A. Singh, Splitstream: high-bandwidth multicast in cooperative environments, in *Proceedings of ACM SOSP 2003*, Oct 2003
21. R. Koetter, D. Karger, M. Effros J. Shi, T. Ho, M. Medard, B. Leong, A random linear network coding approach to multicast. IEEE Trans. Inf. Theory **52**, 4413–4430 (2006)
22. S.Y.R. Li, R. Ahlswede, N. Cai, R.W. Yeung, Network information flow. IEEE Trans. Inf. Theory **46**, 1204–1216 (2000)
23. B.C. Li, Y. Zhu, J. Guo, Multicast with network coding in application-layer overlay networks. IEEE J. Sel. Areas Commun. **22**(1), 107–120 (2004)
24. B. Li, X. Zhang, J. Liu, T.-S. P. Yum, Coolstreaming: a data-driven overlay network for efficient live media streaming, in *Proceedings of IEEE INFOCOM 2005*, 2005

25. Y.T.H. Li, D. Ren, S.H.G. Chan, On reducing mesh delay for peer-to-peer live streaming, in *Proceedings of IEEE INFOCOM 2008*, Apr 2008
26. M. Luby, J.W. Byers, M. Mitzenmacher, A digital fountain approach to asynchronous reliable multicast. IEEE J. Sel. Areas Commun. **20**(8), 1528–1540 (2002)
27. N. Magharei, R. Rejaie, Understanding mesh based peer-to-peer streaming, in *Proceedings of ACM NOSSDAV 2006*, 2006
28. M. Mahdian, K. Jain, M.R. Salavatipour, Packing Steiner trees, in *14th ACM-SIAM Symposium on Discrete Algorithms*, 2003
29. C.K. Ngai, R.W. Yeung, Network coding gain of combination networks, in *IEEE Information Theory Workshop*, Oct 2004, pp. 283–287
30. Y. Qin, X. Xu, J. Zhou, H. Wang, Y. Yang, Joint generation network coding in unreliable wireless networks, in *Proceedings of IEEE GLOBECOM 2011*, Dec 2011
31. S. Seshan, Y.H. Chu, S.G. Rao, H. Zhang, A case for end system multicast. IEEE J. Sel. Areas Commun. (Special Issue on Netw. Support Multicast) **20**(8), 1456–1471 (2002)
32. J. Shi, M. Effros, T. Ho, M. Medard, D.R. Karger, On randomized network coding, in *Proceedings of Annual Allerton Conference on Communication, Control, and Computing*, 2003
33. N. Sundaram, P. Ramanathan, Multirate media streaming using network coding, in *Proceedings of 43rd Allerton Conference on Communication, Control, and Computing*, Sept 2005
34. The Network Simulator NS-2 (2002), http://www.isi.edu/nsnam/ns/
35. A. Vitali, M. Fumagalli, Standard-compatible multiple-description coding (MDC) and layered coding (LC) of audio/video streams. Internet Draft - Network Working Group, July 2005
36. C. Wang, N.B. Shroff, Intersession network coding for two simple multicast sessions, in *Proceedings of Annual Allerton Conference on Communication, Control, and Computing*, Sept 2007
37. Y. Wang, Y. Yang, Multicasting in a class of multicast-capable WDM networks. J. Lightwave Technol. **20**, 350–359 (2002)
38. D. Xu, B. Bhargava, M. Hefeeda, A. Habib, B. Botev, Collectcast: a peer-to-peer service for media streaming. ACM/Springer Multimedia Syst. J. **11**, 68–81 (2003)
39. Y. Yang, A class of interconnection networks for multicasting. IEEE Trans. Comput. **47**, 899–906 (1998)
40. Y. Yang, J. Wang, On blocking probability of multicast networks. IEEE Trans. Commun. **46**, 957–968 (1998)
41. M. Yang, Y. Yang, Peer-to-peer file sharing based on network coding, in *The 28th IEEE International Conference on Distributed Computing Systems*, 2008
42. M. Yang, Y. Yang, Adaptive network coding for heterogeneous peer-to-peer streaming systems, in *The 9th IEEE International Symposium on Network Computing and Applications*, 2009
43. M. Yang, Y. Yang, An efficient hybrid peer-to-peer system for distributed data sharing. IEEE Trans. Comput. **59**(9), 1158–1171 (2010)
44. M. Yang, Y. Yang, A hypergraph approach to linear network coding in multicast networks. IEEE Trans. Parallel Distrib. Syst. **21**(7), 968–982 (2010)
45. M. Yang, Y. Yang, Optimal overlay construction on heterogeneous live peer-to-peer streaming systems, in *The 39th IEEE International Conference on Parallel Processing*, 2010
46. Y. Yang, J. Wang, M. Yang, A service-centric multicast architecture and routing protocol. IEEE Trans. Parallel Distrib. Syst. **19**, 35–51 (2008)
47. Y. Yang, X. Deng, S. Hong, A flexible platform for hardware-aware network experiments and a case study on wireless network coding. IEEE/ACM Trans. Networking **21**(1), 149–161 (2013)
48. R.W. Yeung, S.Y.R. Li, N. Cai, Linear network coding. IEEE Trans. Inf. Theory **49**, 371–381 (2003)
49. Q. Zhang, Z. Zhang, J. Zhao, F. Yang, F. Zhang, Lion: layered overlay multicast with network coding. IEEE Trans. Multimedia **8**(5), 1021–1032 (2006)
50. M. Zhao, Y. Yang, Packet scheduling with joint design of MIMO and network coding. J. Parallel Distrib. Comput. **72**(3) (2012)

Erratum to:

Throughput of Network Coding Nodes Employing Go-Back-N or Selective-Repeat Automatic Repeat ReQuest

Yang Qin and Lie-Liang Yang

© Springer International Publishing Switzerland 2017
Y. Qin (ed.), *Network Coding at Different Layers In Wireless Networks*,
DOI 10.1007/978-3-319-29770-5_2

DOI 10.1007/978-3-319-29770-5_6

The previously published version of the book had an incorrect author credit for Chapter 2.

The correct affiliation of the corresponding author is as follows:

Y. Qin
School of Electronics and Computer Science, University of Southampton,
Southampton SO17 1BJ, UK
e-mail: yqinphd@gmail.com

The updated original online version of this chapter can be found under
http://dx.doi.org/10.1007/978-3-319-29770-5_2

© Springer International Publishing Switzerland 2017
Y. Qin (ed.), *Network Coding at Different Layers In Wireless Networks*,
DOI 10.1007/978-3-319-29770-5_6

Index

© Springer International Publishing Switzerland 2016
Y. Qin (ed.), *Network Coding at Different Layers In Wireless Networks*,
DOI 10.1007/978-3-319-29770-5

Printed in the United States
By Bookmasters